net, in quo terram cum orbe lunari tanquam epicyclo contineri diximus. Quinto loco Venus nono mense reducitur. Sextum deniq; locum Mercurius tenet, octuaginta dierum spacio circũ currens. In medio uero omnium residet Sol. Quis enim in hoc

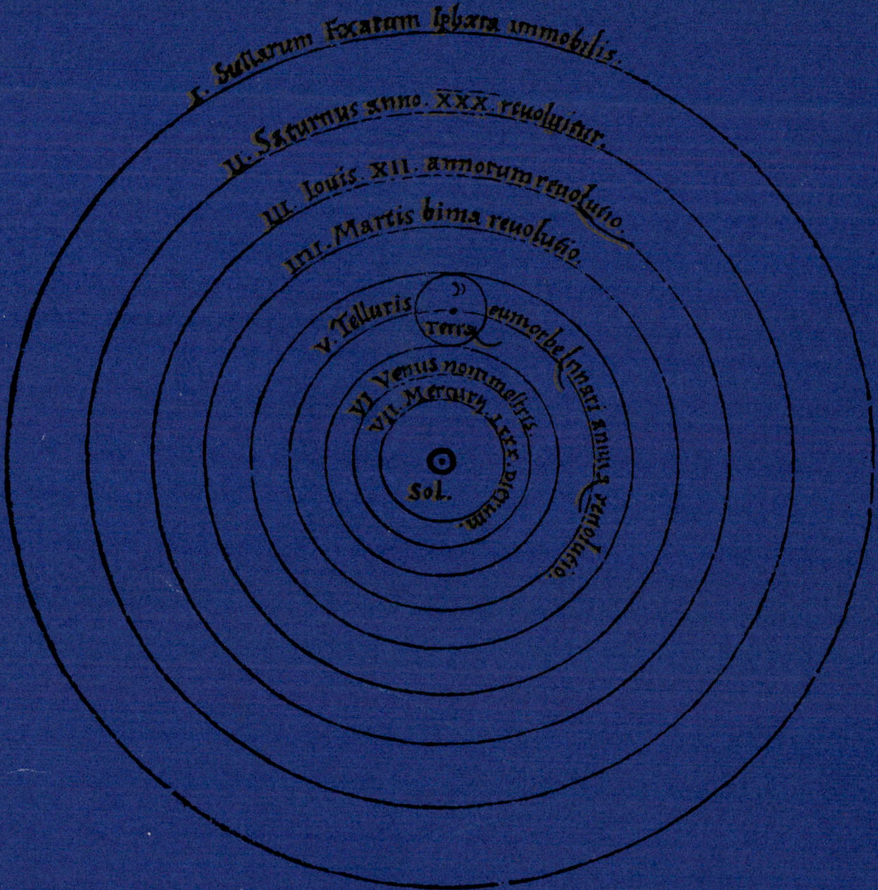


pulcherimo templo lampadem hanc in alio uel meliori loco poneret, quàm unde totum simul possit illuminare? Siquidem non inepte quidam lucernam mundi, alij mentem, alij rectorem uocant. Trimegistus uisibilem Deum, Sophoclis Electra intuentẽ omnia. Ita profecto tanquam in solio re gali Sol residens circum agentem gubernat Astrorum familiam. Tellus quoq; minime fraudatur lunari ministerio, sed ut Aristoteles de animalibus ait, maximã Luna cũ terra cognationẽ habet. Concipit interea à Sole terra & impregnatur annuo partu. Inuenimus igitur sub hac

COPERNICUS QUINCENTENARY

1473—1973

Nicholas Copernicus — the 16th century "Toruń" portrait in The Toruń District Museum collection.

# LAND OF NICHOLAS COPERNICUS

**Text by**
**Michał Rusinek**

**Photographs by**
**Magdalena Rusinek—Kwilecka**

**Translation by**
**A.T. Jordan**

TWAYNE PUBLISHERS, INC., NEW YORK
AND
THE KOŚCIUSZKO FOUNDATION
1973

Graphic Design
FRANCISZEK BARĄCZ

Editors
MARIA PETRI-SZELHAUS
HALINA PRÓSZYŃSKA

Technical Editor
DONAT CHRUŚCICKI

Photos by different authors

| | | | |
|---|---|---|---|
| E. Hartwig | — | pp. | 127/II, 128, 130, 131 |
| C. Kuchta | — | pp. | 2, 6, 61, 86, 88, 89, 127/II, 190, 206/I |
| E. Porębska | — | pp. | 19, 91/II, 132, 142, 154, 160/III, 161/II, 184/II, 188, 189/II, 200/II, 208, 219 |
| J. Tarnowski | — | pp. | 48/49 color |
| CAF | — | pp. | 28, 33, 113, 126/I, 126/II |

The original Polish version ZIEMIA KOPERNIKA
was printed in Poland by
The Sport i Turystyka Publishing House, in 1972

This publication was made possible thanks to a grant
from The Alfred Jurzykowski Foundation, New York

THE LIBRARY OF POLISH STUDIES
Eugene Kusielewicz, **General Editor**
VOLUME II
LAND OF NICHOLAS COPERNICUS

Printed in Poland
by
The Sport i Turystyka Publishing House
1973

## CONTENTS

Nicholas Copernicus, woodcut from the end of the XVI century, called the "Kauffman Copernicus" after the name of its publisher.

# INTRODUCTION

I have tried to sketch the life of the great Polish astronomer against the background of his times, the world in which he lived and carried on his immortal work and the planet which he launched on its orbit around the sun, thus initiating the Copernican era and a revolution in man's concept of the universe. Today the Copernican era glories in the fact that man can travel beyond the earth — in the Cosmos charted by Copernicus and governed, as he put it in his **De Revolutionibus,** "by the admirable harmony of the universe."

On behalf of both authors of this book, the photographer and myself, I extend our sincere thanks to all whose assistance made this work possible — above all to Dr. Eugene Kusielewicz, president of the Kościuszko Foundation in New York; Prof. Joseph Trypućko, director of the Institute of Slavic Philology in Uppsala, Sweden; Prof. Riccardo Lewański in Bologna; the administration of the State Archives in Ferrara; and the translator, A.T. Jordan.

Michał Rusinek

# COPERNICUS' ROOTS AND EARLY YEARS

The Poland in which Copernicus was born and lived was one of the largest countries of Europe, and under the Jagellonian kings extended as far as the sources of Oka and Donets rivers in the east, Brandenburg in the west, the Baltic shore with Courland in the north, while in the south its frontiers reached the Carpathians and the mouth of the Danube on the Black Sea.

Life for the Kopernik family (the Latinized spelling of Copernicus was adopted by Nicholas in accordance with the custom of the scholars of his time) was centered along a path intersecting this great circle—from Wrocław to Cracow, Warsaw, Toruń and Gdańsk, and as far as Frombork at the northern edge of the province of Warmia. The family originated in Silesia, but the name of Copernicus entered history in the north, in a "distant corner of the land" as Nicholas Copernicus described the town of Frombork in a letter to Pope Paul III.

A village called Koperniki still exists, now seven hundred years old, near the county town of Otmuchów in the province of Opole. The forbears of the astronomer had lived in the village and their names are recorded in XIV-th century chronicles as those of townsmen pursuing various trades. One of them was Nicholas, an armorer in Cracow in 1375; another Kopernik was in charge of the Cracow municipal bath-house, and yet another Nicholas Kopernik was master bricklayer in a village near Cracow in 1396. The town records for those years also show the names of other members of the family, such as John, a merchant, and another Nicholas, a master ropemaker. There were among them master craftsmen in several trades, masons, copper miners and dealers in copper.

Some of them, including John, became wealthy merchants through their skills, industry and thrift. We find the astronomer's father, also called Nicholas Kopernik, as a dynamic man of business established in Cracow. His wide interests took him far afield; for example it is on record that he visited Gdańsk in August 1454 as the representative for the merchants of Cracow. He evidently enjoyed their confidence, since he was charged with collecting payment on a large debt, amounting to several hundred zlotys in gold. This was part of a substantial loan made by the citizens of Cracow, including Kopernik himself, to the Prussian League which was loyal to Poland, in support of a military campaign for the liberation of Prussia from the rule of the Teutonic Knights.

Thus Copernicus' father gave financial support to the struggle against the Teutonic Order, which his son continued later when he led the defense of the town of Olsztyn. In the 1460's Nicholas Kopernik, the able Cracow merchant, moved to Toruń and established himself there while still single. Around 1460 he met Barbara Watzenrode, daughter of a prominent Toruń patrician, who was intensely loyal to Poland and not only supported with funds the struggle against the Teutonic Order, but shed his own blood in a battle against the Order near Malbork. He was evidently held in high esteem, for he held judicial office from 1440 and became the senior member of the Toruń bench in 1455, a position he held until his death.

After marrying the daughter of Judge Watzenrode, Nicholas Kopernik prospered; he entered the circle of influential city fathers and when his father-in-law died, he succeeded him on the judicial bench. By 1468 he was the owner of a house at 17 St. Anne's Street, and another in the Market Place, which bears today the number 36.

Recent research established that his wife gave birth to four children at 17 St. Anne's

Street (now proudly known as Copernicus Street). They were: Andrew, Barbara, Catherine and Nicholas, the youngest, born on February 19, 1473.
The future genius spent his childhood there until the age of seven, when his father sold the house and the family moved to the Market Place.
Some Gothic walls are all that remain today of the house in St. Anne's street. It was from its windows that little Nicholas first saw the sun's journey across the sky. The streets he walked as a child are still much the same as then.
What did young Copernicus see around him? What part of the city of his time survived to our day? Toruń is a city where history is ever present, little changed by the centuries. Little Nicholas probably strolled down to the Town Hall with his father, as we can do today. It was then eighty years old, with a Gothic façade designed toward the end of the XIV-th century by the architect known as Master Andrew. On Sundays Nicholas probably attended the church of the Virgin Mary, even older than the Town Hall, as it dates from the XIII-th century. In the streets, he mingled with a multi-lingual throng, for the town was populous (20,000 inhabitants, a very large number for the time) and crowded with merchants and sailors from all over the world. The port of Toruń, on the Vistula river, hosted ships of many flags. Splendid sailing vessels, frigates, brigs, schooners from the Hanseatic cities, ships from Lübeck, London and Stralsund were to be seen at the wharves. Toruń was known as the Queen of the Vistula and the crossroads of trade between western Europe and Poland, Bohemia, Hungary and regions to the east.
The town was rich and lively, with picturesque riverside quays and streets filled with visitors from many countries. There was much to be seen and a chance to rub shoulders with people from faraway lands, to catch the sound of foreign speech.
Young Nicholas' life in such a town could not have been monotonous or dull. The bustling port, with its innumerable contacts with the outside world, was sure to stimulate a boy's mind and arouse an interest in distant, mysterious places.
But Nicholas did not long enjoy family happiness and the care of his devoted father. In mid-1483, when the plague descended on Toruń, the ten year old Nicholas walked behind the coffin of his father, who died in the prime of life.
From then on, the Kopernik family was under the care of Luke Watzenrode, the bereaved widow's brother. Though rather cold and austere, uncle Luke had a better education and a broader knowledge of the world than Nicholas' father, which must have contributed to the boy's intellectual development. Luke Watzenrode was a graduate of the Cracow Academy and had also pursued studies in Cologne and at the famou suniversity of Bologna, where he later sent his nephews Nicholas and Andrew Kopernik. He also knew Rome and was no stranger to the Roman Curia, where he had spent some time. When Nicholas Kopernik senior died, uncle Luke was a member of the court of Zbigniew Oleśnicki, archbishop of Gniezno, a powerful lord who valued the services of the learned Dr. Watzenrode.
While young Nicholas attended school in Toruń and later probably the cathedral school in Włocławek, the family began to disperse. His sister, Barbara, entered the order of the Benedictine nuns, while Catherine married a Cracow merchant.
In the meantime uncle Luke, an older man of earnest disposition, was advancing in his

career. On June 3, 1489, the chapter of the diocese of Warmia elected him bishop, and Pope Innocent VIII soon confirmed the election.

Watzenrode gained in influence and fortune, which he freely used to educate his two nephews. In 1491 Nicholas, then nineteen, and his elder brother Andrew were already students at the Cracow Academy. The clerk entered the name "Nicolaus, Nicolai de Thorunia" in the university register, little knowing that it was to be the most famous of all thus far listed there. A grateful alumnus, Nicholas the son of Nicholas of Toruń, later spoke warmly about the school to which he owed much of his vast knowledge. He did so as a mature man, a scholar familiar with several other universities, but still unaware that he was to bring fame to Cracow and that he would some day come back to the same courtyard of the Collegium Maius in which he had strolled as an undergraduate. He returned as a statue, greeting new generations of students. The Copernicus memorial, by Cyprian Godebski, was erected in the Collegium Maius courtyard toward the end of the XIX-th century, but after World War II was moved to the park near the Collegium Novum.

Let us follow the steps of the thoughtful student in XV-th century Cracow. He strolled along the finest of the streets of the old city, Floriańska, gazing at the towers of the church of the Virgin Mary. The crown of gold did not as yet adorn the belfry, but already then the trumpeter sounded a broken note at the end of every hour, as he does today. High above the twin towers was the sun, still moving steadily from east to west around an immobile earth, obedient to Aristotle and Ptolemy, and the Holy Writ which said: "Let the earth be still for ever, and sun rise and set."

That was also the opinion of the eminent astronomer, master of that marvellous science, Adalbert of Brudzewo, whose lectures young Nicholas was about to attend. To do so he had to cross the Market Square, passing by St. Adalbert's church on his way to the Academy.

But Nicholas Copernicus, the future astronomer, philosopher, mathematician, physician, economist, also had the soul of a poet; he made drawings of the sky, painted his own likeness and even wrote an introductory poem to a book of verse by his friend Dantyszek. Such a young man could not remain insensitive to the beauty and grace of the Market Square of Cracow. Raised in the Catholic faith, he probably offered a prayer at the church of the Virgin Mary, perhaps curious to see if the great altar by Wit Stwosz was soon to be installed.

Nicholas may have uttered there the words he later wrote in Polish in the margin of one of the books he was reading: "Lord help me." After which he then would emerge from the cool shade of the Gothic arches into the blaze of the still restless sun.

This son of a Cracow merchant may have recalled his father as he entered the Cloth Hall, a bustling mart filled with the chatter of the merry wives of Cracow. Young Copernicus sensed the balance of the Market Square with its ring of magnificent patrician town houses, adorned on the south side by the gem of St. Adalbert's church, and the Town Hall tower reaching boldly into the sky. The high towers reaching out to heaven were the points that attracted the gaze of the incipient astronomer.

In daytime he saw plainly how the sun slowly passed the tall towers of St. Mary's and then described an arc above the Square, to sink behind the Town Hall tower. And at night the dark velvet canopy of the sky, the "ninth sphere" according to the learned men of the Church, also moved across the towers, propelled by Cherubim, along the same path as the sun. It seemed obvious, the Market Square was there, the towers immovable, the stones unflinching and the city asleep on a motionless earth, anchored since the dawn of time. Could the young student of Cracow Academy have dreamed then that some day he would halt the crystal sphere, dethrone the archangels moving it and tell the sun to stop in its tracks? Could he have imagined that his feeble human hand would, by the power of the pen, launch the earth into its daily orbit and a trajectory along boundless paths in space beyond the grasp of understanding?

# THE CRACOW ACADEMY

Both the Academy and the city of Cracow enjoyed wide renown throughout Europe. It was the capital of the vast realm of Poland, known for its wealth and Gothic beauty. It attracted artists, artisans, scholars and many foreign merchants. There was no shortage of goods in the Market Square, the Cloth Hall and the many stores. The townsmen had sumptuous town houses and the gentry fine mansions.

Copernicus must have admired both the city of Cracow and the nearby separate town of Kazimierz, on the other side of the Vistula river, accessible by the bridge near St. Andrew's church. The twin cities, separated by the river, were both heavily fortified with massive defense walls and dozens of towers, some of which exist to this day, as for example the Carpenters', the Joiners', the Butchers', the Haberdashers' and the Florian Gate, and by the bewitchingly beautiful bastion outside the city walls — the Barbican. There were innumerable churches, chief among them St. Mary's and the cathedral on the Wawel hill.

There, in the somber cathedral nave, contemporary visitors admired the tombs of the kings of Poland — Władysław Łokietek, Casimir the Great, and that masterpiece of sculpture, the sarcophagus of Władysław Jagiełło. The latest addition at the time was the monumental sarcophagus of Casimir the Jagiellonian, by Wit Stwosz, placed in the cathedral probably around 1491.

Wawel (pronounced Vavel) castle was then a completely Gothic structure, somewhat different from the one we know today. We can visualize it by looking at the surviving Gothic fragments of the castle: the palatium with the Hen's Foot Tower, the walls and ramparts dating from the XIV-th and XV-th centuries, such as the Thieves' tower, the Sandomierz and Lubranka towers, or the Gothic chamber incorporated in the Renaissance section of the Royal palace. The austere beauty of that magnificent hall is highlighted by the single pillar set in the center to support its intricate arches.

The young student from Toruń did not visit, of course, the royal palace, though he may have entered it later, accompanying bishop Watzenrode when he was received by the king. But he was then only nineteen, and on his way to lectures at the Academy, he crossed the Market Square and came to Collegium Maius. The bustle of the market place dies down and all was peaceful and serene in the confines of the Academy. There was some youthful chatter in the courtyard of the Collegium Maius, girded by graceful balconies, but soon all was silent as the students went to their lectures. Copernicus stepped into the Galen lectorium, as it was there that the masters of his favorite subjects, mathematics and astronomy, taught.

At the Cracow Academy young Copernicus acquired a solid grounding in mathematical and astronomical science. In accordance with the rules of the time, he had to study numerous subjects required for a bachelor's degree, starting with rhetoric, poetics and grammar. They were taught from the textbook by Alexander de Villa Dei, written in verse. Later he became immersed in Aristotle, and then in the works of ancient and contemporary mathematicians and astronomers. Eager for knowledge, the young student progressed to scientific books of increasing complexity. Some of them he purchased with his own money and kept to the end of his life in his Frombork library. Most of these books, invaluable for the study of the great scientist's intellectual development, were taken by the Swedes in the course of a XVII-th century invasion and are now to be found at the library of Uppsala in Sweden. These folios testify to the early interest of Copernicus in astronomy and in the works of ancient philosophers, the triumphant rediscovery of which was the dominant theme of the Renaissance. Their pages are evidence of Copernicus' painstaking scholarship and the critical attention with which he studied scientific material.

The marginal notes in his own handwriting reflect the views of the young scholar as well as later of a mature scientist, helping historians of science to reconstruct his life and the evolution of his personality.

Sometimes a single word or a brief sentence in the astronomer's hand reveals the workings of his personality, of his restless mind, ever searching for truth about the nature of the universe.

He sometimes asked himself questions, jotting down such remarks as:

"Is the world eternal?"
"Is the earth naturally located at the center of the world?"
"Is there life on the equator?"
"Are the poles mobile or stationary?"

One could quote many of these random reflections of Copernicus, which he used to note in the course of his readings, especially since they were not concerned with astronomy alone, but dealt with many other areas of science, mainly medicine. They all show great industry and a precise analytical mind, hinting at his doubts concerning the current state of knowledge and revealing how he gradually came to feel that he could undertake a major work of his own.

He acquired that confidence and a solid foundation of knowledge at the Cracow Academy, reputed at the time to be one of the leading universities in the world. That was the opinion of its foreign alumni and the scholars who came from other countries to teach in Cracow. Hartmann Schedel wrote in his **World Chronicle,** a contemporary work "... next to

St. Anne's church there is the university, renowned for its many famous and learned men, where all arts flourish: rhetoric, poetics, philosophy and physics. It is, however, particularly noted for astronomy and in that respect, and as I have heard from many people, there is no finer school in all Germany."

Indeed, Cracow had at about that time, toward the end of the XV-th century, some of the most distinguished scholars of Europe as guest professors, among them the famous Callimachus. Konrad Celtes, the poet laureate crowned by emperor Frederick III, studied mathematics in Cracow. Astronomy was taught at the Collegium Maius by Adalbert of Brudzewo, the internationally famous teacher of Copernicus. Other leading professors included John of Głogów, Michael of Wrocław, Martin Biem of Olkusz, and later Simon of Sierpiec. It was through them and other outstanding professors of the Cracow Academy that Nicholas Copernicus acquired the knowledge of the universe which was then available, based mainly on the writings and computations of Aristotle and Ptolemy. The original text of Ptolemy's **The Great System in Four Books** was not known at that time. European scholars used instead a medieval Latin version of the Arabic translation from the Greek made in A.D. 827 on the order of Caliph Al-Mamun. This translation from Arabic, by Gerhard of Sabionetta, was not entirely accurate or faithful to Ptolemy's original and was known as the **Almagest.**

Even though in a somewhat distorted form, the Arabs made Ptolemy's cosmogony available to the world through the **Almagest.** Besides, Arab scholars of the golden age of the IX-th to XI-th centuries endeavored to improve on the Ptolemaic system, based on the assumption of a stationary earth around which other heavenly bodies revolved. They failed to supplant Ptolemy's system with another, but some eminent astronomers and mathematicians — like Jabir Ibn Afflah and Averroes (known as the Arab Aristotle) — had serious reservations about the hitherto sacrosanct Ptolemaic view of the universe. The doubts aroused in the Arab world by the astronomical tables of Ptolemy inspired Alphonso X, king of Castile, known as the Wise, to take the unusual step of calling a congress of sixty Arab and Jewish astronomers. They were charged by the king with elaborating new principles of the movements of heavenly bodies, that is new astronomical tables.

Copernicus himself noted a comment about king Alphonso on a page of one of the books containing these tables: "Egidius says of him that he was the most benevolent of kings, for he paid a hundred thousand ducats for correcting two folios of astronomical tables." Elsewhere he noted: "Alphonso the astronomer was king of Castile and Spain, whom Haly interpreted as a 'deep source' **(altus fons).**" Alphonso equates himself with an **altus fons.**

The royal astronomical tables actually proved of little value. Nevertheless Copernicus used them, copying whole columns of figures and calculations with extraordinary meticulousness. During his studies in Cracow the principal textbooks used there, in addition to the Alphonsian tables, were the works of the German scholars, Peuerbach and Regiomontanus.

The latter, a pupil of Peuerbach, wrote the astronomical treatise **Tabulae directionum** based on the Ptolemaic system. He also prepared a new Latin edition of the **Almagest,** published in Venice in 1496. This work, entitled **Epytoma Joannis de Monte Regio in Almagestum Ptolomei**, was available to Copernicus during his studies; he must have heard about it from master Simon of Sierpiec, who taught his students about the books and tables of Peuerbach.

But while all these Arab, German or Spanish scientists tried in vain to improve the geocentric system of Ptolemy and rid it of obscurities, none of them attempted to discard it, for they did not think its vision of the universe wrong. The earth remained motionless and the system of Aristotle and Ptolemy, though patched up with some new calculations, remained intact. This was the **ex cathedra** teaching heard by young Copernicus from his university professors. One of them, John of Głogów, must have surely told his students what he wrote in his commentaries: "The earth is immobile in the center of the universe, as philosophers prove by many arguments. The heavens, the purest of all bodies, encircles and bounds the terrestrial dirt."

The young Copernicus attended all these lectures, followed the calculations, copied the intricate mathematical tables of the contemporary authorities of science, made marginal notes on his folios and carried in his mind a growing seed of doubt. He was tormented by questions like: Is the world eternal? Is the earth placed by nature in the center of the universe? Are the poles stationary?

# STUDIES IN BOLOGNA

After completing their studies in Cracow, Nicholas and his elder brother spent some time in the province of Warmia, where their guardian, bishop Watzenrode, who ruled Warmia like a sovereign prince, planned to appoint both his nephews canons of the Warmia chapter.
Watzenrode, a scholarly man, put a high value on education and did not begrudge funds for continuing the studies of his wards. Towards the end of the summer of 1496 the Kopernik brothers were sent to distant Bologna.
At the end of the XV-th century, when Copernicus arrived there, Bologna was one of the wealthiest and most famous cities in Italy. It was called "Bologna the Fat," and people also said of it "Bologna the teacher," for it had the oldest university in Europe.
Young men from all over Europe came to study there, including many Poles, mostly sons of the wealthy nobility. Life in town was expensive and so was study at the university. Besides, the carefree atmosphere of the city invited the enjoyment of life, tempting visitors with entertainment and rounds of pleasure.
The city fathers generously supported their university with a grant of twenty thousand ducats a year — a huge sum at the time. The students, especially the sons of wealthy foreign families, also spent freely on entertainment, servants, dalliance with the daughters of Bologna and feasts combined with intellectual debate, at which adherents of various philosophical schools engaged in fervent, sometimes bitter polemics.
The Kopernik brothers probably shared in the normal life of Bologna students, perhaps less playful than their southern classmates, but surely were enraptured by the exuberant, ebullient lifestyle of an Italian town which they saw for the first time.
They were surely impressed by the architecture of the ancient city. They remembered the towns of Poland, the Toruń, Gdańsk and Cracow churches and cathedrals, but the buildings at the foot of the Apennines revealed other shapes, other colors, with the sun creeping up the red slanting towers of Torre Garisenda and the far taller Torre Asinelli, or building with its glare as one left the dark arcades at the Piazza della Mercanzia, emerging into the square in front of the town hall, surrounded by old houses and palaces.
There was much to be seen. Standing in the Piazza Maggiore, the central point of Bo-

logna, they were surrounded by magnificent buildings. The austere façade of San Petronio, one of the largest churches in Italy, dominated the square. Opposite stood the palace del Enzo, which still stands, and next to it the Palazzo d'Accursio reached into the sky with its finely crenellated profile.

Copernicus must have walked from the square toward the two slanted towers of Bologna, the Asinelli and the Garisenda, on his way to the old university quarter. He could walk all the way under shady arcades, as the rich city provided arches designed to protect its citizens against the rain or the blazing sun. Bologna has to this day over twenty miles of such arcades.

Nicholas was supposed to study law in Bologna, but little is known about this aspect of his education, since he seemed to be more interested in the lectures on philosophy, mathematics, astronomy, medicine and Greek grammar, which he attended at the school of arts.

It was in sunny Bologna that young Copernicus formed a close friendship with one of the professors, a man in his forties, Domenico Maria di Novara, an astronomer and astrologist—there was little distinction between the two at that time. A native of nearby Ferrara, he had previously taught mathematics in Perugia and then in Rome; in 1484 he was appointed professor of astrology in Bologna and was held in high esteem there. His writings, mainly astrological treatises, have not been preserved.

The cordial friendship between the professor and the student was described by Rheticus, a later disciple of Copernicus, who published an outline of his discovery around 1540, several years before the publication of **De Revolutionibus.**

Elsewhere Rheticus wrote: "...In Bologna my doctor, less as a student, rather as an assistant, witnessed the observations of the learned Domenicus Maria and carefully recorded them." In another book, published in 1551, Rheticus noted that: "He was a friend of Sir Maria of Bologna, whose calculations he knew well and whom he helped in his observations."

The German biographer of Copernicus, Hermann Kesten, to whose **Copernicus and his World** we owe the above quotations, observed that "the greatest claim to fame of Maria, who died at fifty in 1504, after teaching for twenty years at Bologna, was the fact that he had Copernicus among his pupils."

Be that as it may, his friendship with Novara kept up Copernicus' interest in astronomy. It was together with Novara that Copernicus made one of his first discoveries, which he later described in **De Revolutionibus.** On March 9, 1497, the two astronomers witnessed the eclipse of the star Aldebaran, also known as Bull's Eye, which disappeared behind the moon. "Waiting for it, we saw the star touching the darker side of the moon and then vanishing between the tips of its crescent...," — thus reported Copernicus.

The young astronomer must have been overjoyed, for he had calculated on the basis of trigonometry the exact moment of that occurrence, and at the same time made an observation which was the first step in disproving the Ptolemaic theory of the moon. According to Ptolemy the moon was supposed to be twice as far away from the earth in quadrature as when full or rising, while the Bologna observation contradicted that assumption.

A year after his arrival in Bologna, in 1496, the twenty-four year old Nicholas Copernicus was appointed, thanks to the support of his uncle, a canon of the chapter of Warmia, filling a vacancy opened by the death of canon Czannow. He signed in the presence of the Bologna notary, Belvisi, an act of acceptance, at the same time authorizing two other Warmia canons to be his deputies for the time being.

The appointment meant a secure financial position. The canons of Warmia, about a dozen in number, were entitled to free living quarters and shared the revenues of the estates of the chapter, which were considerable, as well as they also could have their own farms. The canons could be priests, who celebrated services in the cathedral, but they could also be laymen, without holy orders or theological training. The position was simply a sinecure. It was even possible that none of the canons could perform a mass, which seemed to be the case at the time Copernicus was appointed.

In making their selection, the bishops of Warmia insisted, however, that the canons be educated men and they were consequently required to attend a university in Poland or abroad. The chapter granted a leave of absence of several years for such studies. The records of the Warmia chapter show that Nicholas Copernicus was given in July, 1501 an extension of his leave, after the first one of three years had expired.

The years which Copernicus spent studying in Italy, first in Bologna, then in Padua and Ferrara, coincided with an extraordinary epoch. It was the period just after the discovery of America, when Michelangelo was just starting his work busy at the time in Bologna on a statue of an angel, when Leonardo da Vinci was in his prime and when in nearby Florence the apostle of the new morality, Savonarola, had driven out the Medicis and seized control of the city. Bologna was following with the keenest interest the career of the rebellious friar, as it was precisely from Bologna that he had moved a few years earlier to the San Marco monastery in Florence.

In northern Italy, where young Copernicus travelled, the impact of Savonarola's revolution was tremendous. The humble monk from San Marco, a small and insignificant looking man with an ascetic, intense face, won the support of the masses by the force of his fiery oratory. Savonarola condemned corruption and immorality, he openly attacked the dissolute Pope Alexander VI and publicly burned everything that served vanity and luxury—rich clothes, pictures and jewels, as well as works of art and books by Boccaccio and Petrarch—while urging humility and penance.

Prophet, fanatic and madman, he scorned the Pope's warnings and declined his offer of a cardinal's hat, confident in his mission: the salvation of mankind from sin and a return to strict virtue and austerity. He was convinced that his remarkable power for influencing people, confirmed by all contemporaries, would enable him to overcome all obstacles and institute a reign worthy of Christ, whom he incidentally proclaimed "king of Florence." Five years later the Florentine prophet lost his battle against the Vatican.

Unable to intimidate or bribe Savonarola, Pope Alexander VI decided to destroy him. Intrigues and time did their work. Savonarola was renounced by his earlier supporters, his former fanatical followers betrayed him and delivered him for torture. Reviled by the Florence mob, the once irresistible preacher was sentenced to be hanged and then burned. His ashes were thrown into the river Arno.

All these things happened during the early years of Copernicus' residence in Italy. It was in May 1498 that the body of Savonarola was burned, together with those of two loyal disciples, Domenico Buonvicini and Silvestro Maruffi, who chose to die on the gallows together with their beloved master.

Copernicus, then a student at the university of Bologna, must have followed these events closely. He knew very well what happened to fiery champions of new ideas. It is true that two years later, in the holy year 1500, he would see a commemorative medal struck in honor of Savonarola and distributed in Rome, but it did not wipe out the horror of the execution.

After about four years of study, from 1496 to 1500, Copernicus went to Rome for a year. He had behind him a solid grounding in mathematics, astronomy and Greek, which he probably learned from Antonius Urceus, also known as Codrus. Codrus taught the astronomy of the ancient Greeks and commented on Homer, Euclid, Archimedes and the Greek epistolographers, and he was probably the one who encouraged Copernicus to translate from the Greek into Latin the letters of Theophilact Symocatta, which he later published in Cracow in 1509, dedicating them to his uncle bishop Watzenrode.

# IN THE ETERNAL CITY, PADUA AND FERRARA

Copernicus' stay in Italy coincided with the jubilee year 1500. Alexander VI, the notorious Borgia pope, father of Cesare and Lucrezia, used the sesquimillennium of Christianity as an opportunity for celebrations glorifying his reign.

Thousands of pilgrims, both clergy and laymen, converged on the Holy City. Contemporary chronicles record the attendance of over two hundred thousand of the faithful, which meant that the population of Rome — amounting at the time to about fifty thousand — was temporarily quadrupled.

Copernicus was there, in the throng celebrating the jubilee of Christianity. He did not arrive at the beginning of the holy year, for in March 1500 he was still carrying out astronomical observations in Bologna, but in April he was already in Rome.

He stayed there for about a year, spending more time at work than at the lavish entertainments provided for the visitors from all over Europe. Rome in the year of the jubilee was a babel of all nationalities, filled with people looking not only for religious inspiration, but also for more earthly pleasures provided in abundance by the capital of Christendom.

Rheticus later reported in his book that his master, following the custom of humanists visiting other universities, lectured in Rome on astronomy and mathematics. Rheticus writes that Copernicus "...then aged twenty-seven, lectured in Rome around 1500, as professor of mathematics, before numerous students and in the presence of the leading masters of that science." He also continued his astronomical observations in Rome. Yet, even a studious young man could not fail to be drawn into the whirl of Roman life at the apogee of the Renaissance. After all, Copernicus was only twenty-seven. No wonder that he spent a year in Rome, stretching the leave of absence which the chapter of Warmia had granted him for study at Bologna.

The Rome of the jubilee year under Alexander VI was indeed a fascinating city. The Christian pilgrims flocking to the holy city at the pope's invitation were shocked to discover that it was the scene of violence, murders, orgies and every kind of depravity.

The rapacious pope sold the dignity of cardinal to the highest bidders, and his son Cesare terrorized the city. "There was hardly a night without four or five murders, and there were bishops and prelates among the victims. In the morning of May 27, 1500, Romans saw eighteen corpses hanging from the Sant' Angelo bridge." Soon afterward, Alfonso Bisceglie, the husband of Lucrezia Borgia, the pope's daughter, was killed by thugs sent by Lucrezia's brother Cesare. The Vatican chronicler, Onofrio Panvinio, reports that Alexander VI had caused three cardinals: Orsini, Ferrerio and Michiel, to be poisoned and "would have done away with all the rich cardinals and prelates to seize their wealth, if death had not caught up with him."

Young Copernicus had much to see in Rome. He could not fail to notice that some were praying and observing the jubilee of Christianity, while others fought each other for advancement and riches, while indulging in every vice and villainy.

While in Rome, Copernicus probably studied ecclesiastical law at the Vatican Curia, as was the custom of canons visiting the Holy City. His brother Andrew had received permission for such study from the Warmia chapter, according to documents preserved to this day.

In the middle of 1501 Copernicus was back in Frombork and he appeared together with his brother Andrew before the diocesan chapter, asking for permission to visit Italy again for further studies, this time in medicine. The chapter, no doubt influenced by bishop Watzenrode, granted his request while expressing the hope that canon Copernicus may be "useful as a physician to the most reverend head of the diocese and the gentlemen of the chapter."

It was not an empty phrase, for Copernicus actually filled the position of physician to the chapter for many years afterwards.

After having received in July, 1501, permission for another trip to Italy, the Copernicus brothers started on their southward journey in the beginning of September, accompanied by an older friend, Bernard Sculteti.

It was a journey that usually took many weeks, sometimes months, depending on the route chosen. The Copernicus brothers went by way of Wrocław, where Nicholas was canon of the church of the Holy Cross, which he assumed by proxy while studying in Bologna. He surely must have stopped for a few days in Wrocław, to rest after many days of travel, and to see the ancient city where he held the office of canon. Wrocław had belonged already in the X-th century to Polish princes of the house of Piast, and at the time of Copernicus, was one of the major cities of Europe, a crossroads of trade routes from the south to the north and from the west to the east.

Copernicus, a man of sensitive perception, with aesthetic appreciation finely developed in Italy, no doubt admired the architecture of Wrocław, the fine town houses of the wealthier burghers, the Gothic town hall, dating from the XIII-th century, with its magnificent portal, the XII-th century Romanesque cathedral and the old Salt Square. He must have looked with particular interest at such churches as the XIV-th century St. Elizabeth's, or the tiny St. Ido's, founded in the XIII-th century by a Polish nobleman. The church of the Holy Cross, with which his canonry was associated, was already then two centuries old. Today it is particularly dear to Poles because Polish services were held in the lower part of the church until the end of the XIX-th century, when Wrocław, then a part of Prussia, had been subjected to Germanization.

After stopping in Wrocław, Copernicus and his companions started on the next stage of their journey, with Padua as their destination.

Crossing the mountains was the most strenuous part, which the travelers had to cover mostly on foot, with the scenic beauty of the Alps their only reward. There may have had starlit nights in the mountains, bringing the astronomer closer to the heavenly bodies with which his mind was absorbed.

Padua, surrounded by defense walls, moats and bastions, was a quiet and industrious city. About 25 miles distant from Venice, it belonged to the Venetian Republic, which took very good care of the university there. It provided it with a new building in 1493, as the Copernicus brothers probably knew, since Bologna was not very far from Padua. Then as today, Padua's center was at the square delle Erbe, dominated by the vast XII-th century palace della Ragione, rising in tiers of galleries with arches. The shady arcades and the bustling market under them may have reminded Copernicus of the familiar Cloth Hall of Cracow. The canon of Warmia probably also visited the old basilica del Santo, a huge Romanesque-Gothic structure adorned with an assortment of Byzantine domes.

The various faculties of the university of Padua had been previously dispersed in many buildings scattered throughout the city, mainly around the San Biaggio street, where a tablet attests to the fact today that the school of liberal arts and the school of medicine was once housed there.

Since 1493 the Palazzo del Bo (palace of the ox) became the university center. Today we find only fragments of the splendid building which was new when Copernicus worked there. There is an old tower, galleries and lecture halls, which he no doubt frequented. The university of Padua, as well as that of Bologna, had many Polish students, recalled today by over a hundred Polish coats-of-arms seen in some of the lecture halls. There were also Polish professors and in the academic year 1563–64 Jan Zamoyski was the rector of the university of Padua. Today we find there not only coats-of-arms, but also commemorative tablets and portraits of many eminent Poles, including Jan Kochanowski, Clement Janicki, chancellor Zamoyski and, of course, Nicholas Copernicus.

There is the podium from which Galileo spoke, championing the heliocentric theory of the universe and thereby endorsing the great work of Copernicus, until he was placed on trial by the Inquisition.

The university of Padua enjoyed renown in many fields of learning, but it was particularly famous for its teaching of medicine.

The curriculum of the faculty of medicine included three years of theoretical studies, the dissection of corpses, and one year of practice under the guidance of a professor. At the time when Copernicus was enrolled there, the faculty included such men as Bartolomeo de Montagnana, author of works on contagious diseases; Gabriel Zerbi, anatomist and philosopher, who published in 1502 the **Anatomy of the Human Body**; Antonio Gazzi, author of the **Wreath of Flowers, or the Maintenance of Health,** an outstanding scholar who was at one time physician to Sigismund I, king of Poland.

Copernicus was meticulously conscientious in every task he undertook and he studied

medicine with the same dedication, perhaps even with enthusiasm. This is evident from his notes and the marginal jottings which he had a habit of writing on the books he read. Among those hand-written notes, still preserved, there are some observations, thoughts or recipes that are in no way startling, and others which sound comical today. This does not detract anything from the greatness of the scientist's mind, but merely reflects the medical standards of his time.
Here are some of the innocuous remarks:

"Those who inherited their condition from their parents can seldom be cured, but bear their sufferings with more patience."
"Washing one's hands with hot water before a meal, and cold after, is good for digestion."
"The juice of birch leaves helps in any condition."

Among the more imaginative recipes recommended at the time, Copernicus copied the following:

"Take two ounces of Armenian clay, half an ounce of cinnamon, 2 drams of citwar; two drams each of chicken grass, dyptan and red sandalwood; a dram each of shavings of ivory and iron; two scruples each of ashes and sour rumman; a dram each of lemon peel and pearl; a scruple each of emerald, red hyacinth, sapphire; one dram of the bone from a deer's heart; a scruple each of sea locust, the horn of a unicorn, red coral, gold and silver in flakes; half a pound of sugar, or such an amount as one usually buys for one Hungarian ducat."

In times when astrology was regarded as a science and health was believed to be dependent on the conjuncture of stars, this was a valuable prescription, even if untested. Copernicus evidently never prescribed or concocted it, if only because some of the ingredients could never be secured.
Copernicus was not wholly absorbed by his medical studies, as his mind was still wrestling with the mystery of the structure and motion of the universe. He encountered others with similar interests in astronomy and philosophy. Girolamo Fracastoro, the future reformer of the calendar, was in Padua at the time, as were Luke Gaurico and his brother Pomponius, the former an astronomer and the latter a poet and aesthete.
The two years which Copernicus spent in Italy after 1500 were a particularly active period in his life. Between the fall of 1501 and the end of 1503 he was engaged in many pursuits. He studied medicine and practiced it, a necessary condition for a degree; he continued his astronomical observations and participated in debates, while at the same time perfecting his knowledge of Greek and preparing for an examination in canon law.
In the spring of 1503 he went to Ferrara to appear before the board of examiners and secure a doctoral degree.
Ferrara was at the time one of the wealthiest cities in Italy, also noted for the fact that on the throne of Ferrara was the beautiful Lucrezia Borgia (daughter of pope Alexander

VI), who had married in 1502 the duke of Ferrara, Alphonso. The city was not only rich, but populous, with as many inhabitants as Rome. It also had a reputation for a gay and abandoned life, perhaps associated with the "Pulcherrima virgo" Lucrezia, duchess of Ferrara.

The castle of the dukes d'Este dominated the center of Ferrara at the time, with its forbidding red walls, four tremendous towers, drawbridges and deep moats. Its massive medieval walls are today a tourist attraction, but then they were tangible proof of the power of the rulers of Ferrara.

The battlements of the castle inspired awe, but the vast square around the magnificent Romanesque-Gothic cathedral must have been a cheerful place, filled with a colorful crowd. But Copernicus did not come to Ferrara for recreation. He swore an affidavit that he had completed the required studies in Padua and then had to defend his doctoral dissertation before the assembled faculty.

The ordeal was crowned by the rich pageantry of the ceremony of the formal awarding of the doctor's degree. It included the presentation of the doctoral insignia, with the ring and the parchment diploma, the placing of the doctor's cap on the candidate's head and appropriate speeches. The diploma, a copy of which is still kept in the archives of the university of Ferrara, reads:

"The venerable and learned Sir Nicholas Copernich de Prussia, canon of Warmia and scholar of the church of the Holy Cross in Wrocław, who studied at Bologna and Padua, was confirmed in canon law without opposition and granted a doctor's degree by the Vicar Superior ...."

It is not known how long Copernicus remained in Ferrara and why he chose to receive his doctorate there rather than in Bologna or Padua. His biographers are inclined to think that the motives were financial, as the fees at the university of Ferrara were much lower than in Bologna or Padua.

He did not stay in Ferrara very long, since he still had some medical studies to complete in Padua, but neither did he remain there long, as a chronicler noted his presence at the Council of the Estates of Prussia in Malbork and Elbląg at the beginning of 1504.

# RETURN TO WARMIA

The exciting period of foreign travel, a memorable time in every young man's life, was over and Copernicus returned to Poland, after spending in Italy nearly ten years with brief intervals.

He came back rich in learning and in wordly experience. He had seen the Italian cities with their plebeians and patricians, the students and the dissolute life prevalent at both the secular and the ecclesiastic courts of Italy. He encountered some of the greatest minds of his time, the scholars of three of the leading universities, but he also witnessed the ignorance of some dignitaries and magnates. He returned to his native country not only as a scholar, but also as a man of the world with wide experience in human affairs.

During the next six years he served at the court of his uncle, bishop Watzenrode, in Lidzbark in Warmia, where he was the bishop's confidential aide, inspector of the cities of Prussia and Warmia, entrusted with financial missions and other duties. We find in the records of that period such documents as for example one attesting that "Balthazar Stockfish received two hundred eighty five marks from Messrs. Lossainen and Copernicus, paid them into the treasury of the Chapter, and duly accounted for that sum."

Lidzbark, then the seat of the bishop who was actually the feudal lord of Warmia, is now a quiet small town on the river Łyna, which circles the bishop's castle. The hilly country around the massive walls of the castle, still in excellent condition, was formerly a vast forest full of wild game, including bison, boar, lynx and bear.

The high-rising battlements of the castle dominated the whole region and held enemies at bay. Even now the massive towers at the corners of the fortress inspire respect and hint at the defense potential of the ancient structure. As we enter the castle courtyard, we are bewitched by the double-tiered Gothic galleries resting on graceful arches. We can visualize Copernicus strolling under these arches with his uncle, the bishop. The arcades already existed at that time, as the castle was erected around 1400, but some portions were added later, as for example the chapel built in 1500. One of the finest interiors in the palace is the great reception hall with a characteristic star-shaped arched ceiling.

It was in that magnificent palace that Nicholas Copernicus lived for many years the life of courtier, physician, counselor and financial confident. No one knows how much time he devoted during these years to pondering over the mysteries of the universe, how many astronomical observations he made from the towers of Lidzbark castle, though it is known that in 1511 he observed a total eclipse of the moon, and in the following year he carried out observations of Mars.

He certainly did not abandon his scientific zeal, nor did he give up the hope of solving the mystery of the universe, but Lidzbark was not the scene of his major astronomical discoveries. He was burdened there by too many other duties, among them medical practice, as he was the bishop's official physician. The care of his aging and ailing uncle compelled him to pursue further medical studies and to read contemporary treatises on the subject, jotting down on the margins his personal notes, as usual.

As the trusted aide and personal physician of the ruler of Warmia, Copernicus often traveled with his master all over Poland, particularly in Warmia and Pomerania.
He attended the session of the Estates of Royal Prussia, convoked by Watzenrode on January 18, 1504, in Elbląg. It resolved to swear an oath of allegiance to the king of Poland, Alexander Jagiellon. In the same month of January Copernicus probably attended together with bishop Watzenrode the wedding of the latter's niece, Cordelia Allen, with the banker Reinhold Feldstedt.

When King Alexander visited Toruń in May 1504 and then went with his court to Malbork, Elbląg and Gdańsk, he was accompanied by the bishop of Warmia and in his entourage no doubt, Copernicus. The prince-bishop's nephew and confidential aide must have visited these cities on many other occasions, as Elbląg and Gdańsk were both busy ports, and centers of trade and political activity.

Elbląg, built on the ruins of an ancient settlement of the Prussian tribe, which the Teutonic Knights had exterminated, was prospering at the time thanks to its overseas trade and to its incorporation into Poland in 1466. It was a fortified city with many defense bastions and gates, some of them still in existence, but also such gems of architecture as St. Mary's church, or the St. Nicholas church with its nearly three hundred foot belfry. The old city walls with the Gothic Market Gate are another relic of the past.
Malbork, the main fortress of the Teutonic Order in the former territories of the Prussian tribe, was between 1309 and 1457 the seat of the grand master of the Order. History notes many battles in the course of which the fortress was besieged by Polish forces in 1410, 1457 and 1460, when it was finally captured. In Copernicus' lifetime Malbork was the capital of the province of Pomerania. He knew well that vast, grim castle, awesome symbol of Teutonic power. Built on the river Nogat, vast in scale, it was the repository of the Order's wealth as well as the center of its military might. There is the Gothic middle castle, the upper castle, and the medieval town hall in the lower castle, with the chapel of St. Lawrence, the armory and the foundry.

Copernicus undoubtedly visited Gdańsk often, where he had relatives and much official business. Gdańsk was one of the wealthiest cities in Poland, a center of the export of grain and lumber, as well as the import of foreign goods. It also had industry and notably

iron workshops, armorers and artisans making the finest furniture. The XV-th and XVI-th centuries were a period of great prosperity for Gdańsk, which obtained at that time royal charters granting various privileges to the city.
Gdańsk had many fine old buildings, such as the XIV-th century Great Mill, or the Old Crane on the Motława river, dating from the same time, and the magnificent Artus Hall, which later received a new Renaissance façade. There were many splendid Gothic churches, among them St. Mary's, St. John's and St. Nicholas's. The city walls were reinforced by Gothic bastions and gates, such as the Corner Tower, the Swan Tower, the White Tower and many others. At the turn of the XV-th and XVI-th centuries many new buildings were erected. It was then that the town hall was enlarged and embellished with Polish eagles, the Judges' Hall was built, as well as new defense towers on the Motława river, called the "Milk Jugs" because of their shape (1517—1519).
Amidst these red brick buildings and along the quays of the harbor thousands of sailors and merchants of many nations, Polish squires with their crops, and townsmen from other Polish cities milled in a busy throng, filling the taverns by the Long Market and the Royal Way, now the most historical parts of the city.
Soon after King Alexander's visit in Toruń Copernicus attended the conference of the Estates of Royal Prussia at Malbork and even acted there, at the request of the primate of Poland, as translator of important documents into Latin. It seems very probable that he also visited Cracow at about that time, recalling there the days of his youth, as he must have accompanied the bishop in January 1507 at the particularly solemn coronation of Sigismund I.
It was a very important year in the great astronomer's life. Students of his work, notably Ludwik Antoni Birkenmajer, assert that it was then that Copernicus wrote his **Commentariolus** [Little Commentary] a work which was never published, in which he outlined a hypothesis of the movements of heavenly bodies.
The original manuscript was lost, but XVI-th century copies reveal that the **Commentariolus** advanced the heliocentric theory of the universe, which was later developed and mathematically explained in the great **De Revolutionibus.**
The wide range of the knowledge and interests of Nicholas Copernicus was as astounding as his industry. At about the same time, in 1509, he published in Cracow a Latin translation of the Greek book: **Letters, Mannered, Idyllic and Amatorial of Theophilactus Symocatta the Scholar,** and a few months later, in 1510, he produced the **Map of Prussia.** It was lost, but it evidently had considerable political importance, as Copernicus delineated on it some territories disputed between Poland and the Teutonic Order.
The story of the map is mysterious and obscure. It must have been very unfavorable to the Order, as Hans Lilienthal, also known as Fabian of Lezany, tried to steal it from Copernicus' house at the prodding of the Teutonic Knights. This attempt at espionage was reported by Lilienthal in his letter to the chancellor of the Teutonic Order, John von Schonberg. He wrote: ".... I searched all the rooms, but could not find the map, so I assume that he (Copernicus) may have taken it with him, or hidden it away."
Lilienthal had written previously to the chancellor of the Order that he would try to secure the map, if only he could get the key, and in another letter he expressed the hope

of getting "the whole map": "Ich werd hoffe ich gantcz dy mappa obirkommen, solt ich och eyn pferd daran wogen."
The letters of Fabian of Lezany found in the archives of the Teutonic Order reveal the role he played as a spy for the Order, but Copernicus was, of course, not aware of his duplicity. Actually he and the perfidious Fabian went together to Olsztyn as agents for the Warmia chapter to collect the sum of 238 ducats, which they received on January 1, 1511.
Worse yet, a year later, after bishop Watzenrode's death, Copernicus participated in a session of the Warmia chapter, at which the same Fabian was elected bishop. Together with the other canons, Copernicus signed the documents authorizing Tideman Giese to act as the chapter's plenipotentiary in assuring the ratification of this election by the pope.
The passing of bishop Watzenrode on March 29, 1512, permitted Copernicus to leave Lidzbark castle in April. He moved to Frombork, where he spent the rest of his industrious life, with only a temporary visit to Olsztyn.

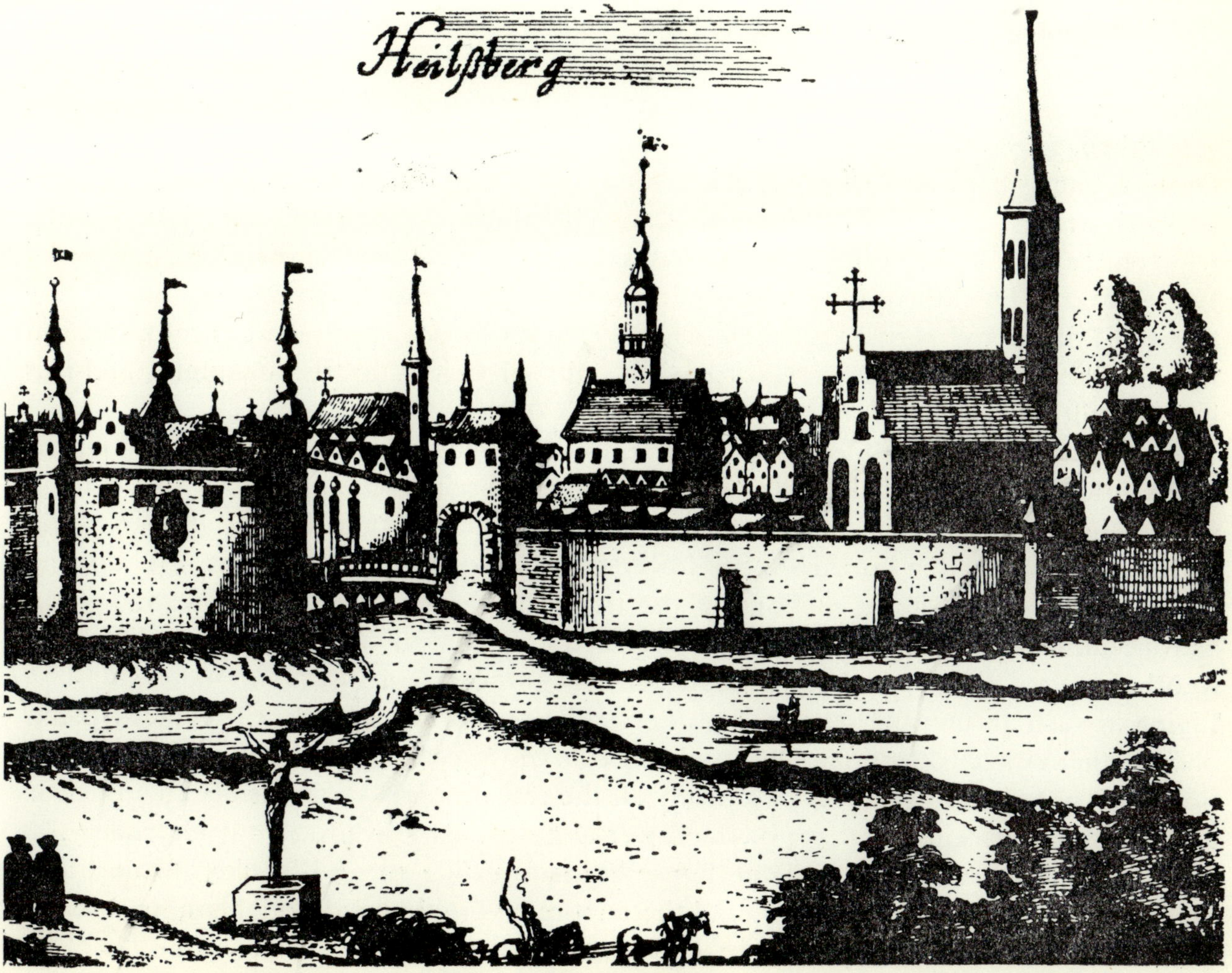

# THE SKY OVER FROMBORK

Frombork, the seat of the Warmia chapter, was in Copernicus' time a fortified town on a hill by the Vistula delta, a particularly well protected site. A ring of red brick defense walls, with bastions and towers, guarded the Frombork cathedral, a magnificent building of impressive proportions.

The Frombork cathedral was built in 1329 and in subsequent centuries it was expanded and sometimes damaged, as for example in 1626, when the Swedish invaders demolished the interior of the church and looted its archives, including Copernicus' library, which was taken to Sweden.

Despite later alterations, the cathedral kept its original beauty. There are the star-shaped arches, the ancient murals, the old altars — especially the late Gothic triptych, carved at the turn of the XVI-th century, reminiscent of the Cracow altar of Wit Stwosz. The cathedral is also noted for its organ, dating from 1684, one of the finest in Poland and perhaps in the world. The first organ was installed in the cathedral in the times of Copernicus; according to bills preserved in the chapter's archives it was built by "master John, organ maker."

The cathedral was surrounded by the curiae, the residences of the canons of the chapter. The whole compound, including the belfry and the cathedral was encircled by the defense walls, reinforced by seven towers, some of which are there to this day.

Copernicus lived in the north-western tower guarding one of the main gates. The living quarters were in the lower sections of the tower, while its upper section served as an astronomical observatory. The gallery attached to the wall of the tower and leading to other towers was particularly suitable for that purpose.

Living in such an unusual setting, high above the rest of Frombork, Copernicus could observe the sky in all directions. Only the steep roof of the cathedral slightly obscured a portion just over the horizon, but practically all the rest of the sky was visible on any clear night from the various windows of the tower and especially from the outside gallery.

Copernicus may have also used the house he had outside the walls, like all the other canons, but he could hardly have found there more favorable conditions than in his tall tower.

The tower became the place where he spent most of his time and it was there that he conducted his scientific work, in calm and solitude, as a free spirit. He wrote once that "astronomy is the science worthy of a free man."

Even then, however, he could not devote himself entirely to science. He had certain duties as a member of the chapter, attending its meetings and allocating land to farmers settled in Warmia. There are in Copernicus' personal notes many references to trips to various villages in Warmia, such as Lajsy, Radziejowo, Naterki, Myki, Jankowo, Tomaszkowo, Gutkowo, Woźno and tens of others. He also noted the names of the peasants he helped to settle, for example Bartosz in Piuski, Polonus in Slupy, Stanisław Czechotzinski in Likusy, etc.

All these activities distracted him from his scientific work, concentrating the search for the truth about the structure and the motion of the solar system. As his biographer M. Adam put it in 1706, he had a "consuming passion for truth." It is astounding that he managed, in that quiet backwater, isolated from the world and without any assistance, using only the most primitive instruments, to carry out the complex and accurate mathematical calculations leading to his epoch-making discovery.

We said "primitive instruments" — actually an understatement. What were the instruments used by the "Sarmatian astronomer" as Melanchthon called him? The principal one consisted of three wooden boards tied together according to his own design, with a scale drawn in ink along them.

Later, after Copernicus' death, the great Danish astronomer Tycho de Brahe wrote the following moving words about that parallactic instrument: ".....with these light rods, simply fashioned, he managed to bring down the stars and reach heights never attempted by any mortal since the beginning of the world."

Even the simplest telescopes did not exist at the time of Copernicus' astronomical activity. The first telescope was aimed at the sky by his successor and admirer, Galileo. Copernicus, like all the other astronomers of his time, had to make his observations with the naked eye. Writing to Pope Paul III about the "distant corner" where he worked, he meant the geographical distance of Frombork from Rome, not any lack of scientific facilities in a backward country.

After travelling extensively, especially in Italy, and knowing well the whole of Poland, with its fine cities, magnificent castles, palaces and schools, Copernicus certainly did not at any time feel that his country was in any way second to those of the rest of Europe. Poland was at that time one of the leading powers of the continent, with vast territories and a strong political and cultural position.

Poland was at that time held in high esteem by other nations. Here is what Erasmus of Rotterdam wrote in the first half of the XVI-th century in a letter to the Cracow banker Decius (according to B. Leśnodorski):

"I congratulate the Polish nation, where arts, science and all things most removed from barbarism now flourish, that it can now compete with the foremost and most civilized nations in the world."

And in 1541, when **De Revolutionibus** was not yet known to the world, though there were some advance announcements of the Copernican discovery, the Flemish scientist Gemma Frisius wrote: "It seems that by some dispensation of providence the Muses, after leaving the spring of Pegasus, journeyed to Sarmatia, attracted either by the land itself or by the genius of its people... Not to mention the others, Urania settled there and found new admirers, bringing us a new earth, a new Phoebe, new stars and even a whole different world." The Italian humanist Paolo Manuzio declared that "among the transalpine nations, none equal Poland in her devotion to science."

* * *

Let us now travel back in time, to that "distant corner" near the Baltic shore, five centuries ago.

There, in the red brick tower by the cathedral, in a backwater far away from the intellectual and political centers of Poland and Europe, the mind of the lonely genius wrestled with the dilemma which had been tormenting it since early youth.

He had for support the wide knowledge garnered at the greatest universities and a painstakingly assembled library. It was precisely his intimate knowledge of the whole of contemporary science that incited him to challenge it in heroic battle, for its doctrines left him unsatisfied and still groping for the truth.

For thousands of years, since Aristotle—the greatest authority of the time—and Ptolemy, the other ancient master of cosmology, through the teachings of the later fathers of the Church, mankind has firmly believed that an immovable earth was at the center of the universe, with the moon, the planets, stars and even the sun itself steadily rotating around it. Only man is granted the privilege of standing on stationary earth and watching the heavenly bodies circle the globe.

How could the lonely scientist, poring over Greek folios and the works of the Church fathers, observing the sky with the naked eye from his aerie, challenge the accumulated wisdom of thousands of years?

Some Church fathers such as Lactantius or Cosmas did not even give credence to Aristotle and Ptolemy, who held the earth to be spherical in shape, but tried to conform to the Holy Writ by asserting that it was flat like a pancake, or even shaped like a tabernaculum. Other wise men of the time had recently claimed that if the earth was indeed a sphere, Columbus could not sail to the Antipodes, for even if he got there, he could never come back uphill.

It was there, in Frombork, that Copernicus wrote in the **De Revolutionibus** that Lactantius "..... a poor mathematician, talks about the shape of the earth like a child, deriding those who described it as a sphere."

Copernicus could easily dismiss such naive ideas, but it was not as easy to cope with the well argumented theories of the classical masters of philosophy, the great mathematicians, or the later advances of the Arab scientists, mostly contained in the adaptation of the Ptolemaic cosmology known under the name of **Almagest** or its medieval versions. That philosophy, endorsed by the fathers of the Church, assumed the earth to be a sphere,

Frombork. The eastern bastion: The Copernicus Tower, part of the fortifications surrounding the Cathedral — built in tbe second half of the 14th century.

with the rest of the universe divided into concentric spherical layers rotating under the control of angels of various ranks.

The first sphere or the first layer of the sky around the earth, including the orbit of the moon, was operated by angels; the second sphere, comprising the planet Mercury, was controlled by archangels; the next sphere, that of the planet Venus, by even higher spiritual forces, and so on.

The ninth and last sphere encircling the earth, a crystal ball studded with stars, was the end of space, with only the throne of the Lord, guarding the whole of the universe.

Aristotle had devised over fifty such spheres rotating at different speeds, as this fitted his calculations of the motion of planets and stars in the sky. At the time of Copernicus, the universe was supposed to be divided — in accordance with Ptolemy — into nine spheres, all rotating at tremendous speed around the center of the universe, that is earth.

The nine heavenly spheres between man and the divine throne had an equivalent in nine circles of hell filled with sinners and devils of various ranks, while at the very bottom of the pit, in the center of earth, ruled Lucifer the Prince of Hell.

Modern man may find it very easy to laugh off such a concept of the universe, but Copernicus, born and raised in an era when religion and science were one, could not dismiss it lightly. As to the actual rotation of the spheres, it was mathematically confirmed by Aristotle himself. The movements of the heavenly bodies in their orbits around earth were calculated with some accuracy and the discrepancies were dealt with by introducing additional orbits, epicycles, and then more epicycles. The basic assumption of such a cosmology was that earth was stationary and all the other planets and stars moved around it.

One stands in amazed awe before the incredible courage and force of character of the hermit from the Frombork tower. Going against all accepted doctrine, and against the authority of the ancients — which carried tremendous weight at the time — he directed the power of his mind to the overthrow of the whole system, while realizing at the same time that, as a Christian, canon of Warmia and servant of the church, he was challenging the tenets of the faith by confronting his new truth with dogma sanctioned by many generations.

* * *

He devoted all the time that the duties of canon and physician left him to observing the sky. He watched the stars and he concentrated his brilliant mind on the riddle of the universe. He made some additions to his library and he read the works of astronomers, mathematicians and philosophers, turning back to Aristotle, Ptolemy, Cicero, Lucretius, Virgil; he studied again the **Almagest** and the works of Vitruvius, which dealt mainly with architecture, but also referred to the heavens, the stars and the atom.

He was a critical reader, accepting the assertions of some authors and rejecting those of others, in terse marginal notes.

The future discoverer of the true structure of the universe increased his already vast knowledge, and with it his doubts of the validity of the science of his day. Some books gave him clues encouraging the protracted search, others he discarded with a smile.

But the sky spread wide above the cathedral belfry, and over the nearby Baltic sea was the most instructive book of all from which the "Sarmatian astronomer" learned more than from the folios of the classics. He watched the movement of the stars, of the moon and the sun, rising every morning over the sandy dunes of the coast and sinking every evening in the west. He knew that the human eye is susceptible to error, that its action is dependent on the observer's position and motion. Already an ancient poet said: "We set sail and the land with its cities recedes into the distance."

No one will ever know the exact moment when his mind was flooded by a new vision of the world, reversing the entire system, but he must have carried the germ of the idea for a long time. It was at first only a hypothesis, an abstract theory, but he wanted to support it by mathematical calculation, to make an incontrovertible statement based on rigorous logic, a true work of discovery.

It was an age of discovery, stimulating to a creative and active mind. Copernicus knew that "under the authority of the Spanish and Portuguese rulers Amerigo reached new lands." He did not mention Columbus, whose achievement was dimmed by that of Amerigo Vespucci. Yet he knew about the discovery of the new world and perhaps was inspired by it. There are some analogies between the two explorers. Like the fearless navigator, who sailed half way around the world though warned by wise men that he could never return "upward," the explorer of the universe set the sails of his lonely mind for a more distant voyage, to discover the sun around which the earth whirled in orbit.

The work in which he set down his observations, calculations and conclusions took him a lifetime to prepare; it assumed its final form in Frombork, where he gazed endlessly at the sky.

# THE TREATISE ABOUT SOUND MONEY

Throughout his life Copernicus gave testimony that he was a loyal subject of the Polish crown, and that the Polish soil was loved by him. He was particularly attached to the northern provinces in which he spent most of his life, to the land of Warmia where he lived, and it was here, in Frombork, that he made his systematic celestial calculations. From his notes we see that time and time again, sometimes within the span of the same month, he observed from his Frombork tower the planets Venus, Saturn, Mars, Jupiter or again the conjuncture of Saturn with the moon.

He was so absorbed by observations, studies, and mathematical calculations that it is hard to see how he managed to find time for his administrative duties and the writing of economic analyses, such as his famous treatise on sound money. The magnitude of his achievement in astronomy inevitably relegated to the background the other activities of that versatile genius, among them his work in the field of monetary theory.

In those days the question of monetary reform was a weighty problem throughout the whole world, flooded as it was by currency of various sorts, and particularly in Poland — which was a big country then and included in the orbit of brisk foreign trade funnelled through Gdańsk — reform became an urgent postulate. Copernicus was fully aware of the importance of a stable currency as a condition for the development of trade between na tions. The situation was deteriorating rapidly. Shortsighted monarchs and those operating mints on their behalf made quick profits by issuing debased coinage, that is reducing the gold or silver content of coins. Sound money disappeared from circulation since people naturally hoarded it. The law which states that bad money drives out the good is generally ascribed to Gresham and bears his name, but it was actually formulated by Copernicus much earlier.

Copernicus wrote his treatise about sound currency for the provincial councils of Royal Prussia and the Polish parliament, to protect them — and notably his own province of Warmia — against the evils of a debased currency. He advocated a uniform currency for the whole country, minted in accordance with standards set by the government, criticizing the existence of several separate mints. In northern Poland alone there were mints in Gdańsk, Toruń and Elbląg, each city producing its own coinage.

Copernicus argued that sound currency was indispensable to a country's prosperity: "nations with sound money have the best artisans, works of art and an abundance of all goods."
Debased money, on the other hand, is a calamity: "Among the disasters that bring the downfall of whole states, whether realms or republics, four are the most dangerous: domestic dissension, high mortality, failure of crops and unsound money... The last of these, the debasement of money, is perceived only by a few, and solely the more serious minds, as the decline of a nation under its influence is not immediate, but gradual and imperceptible."
King Sigismund the Old was endeavoring at the time to establish a uniform currency in Poland and in Prussia, where several mints were operating. Copernicus strongly supported the proposed reform: "Unless prompt action is taken, Prussia will soon have money containing nothing but copper. Then all trade with other countries will vanish. Will any of the foreign merchants sell their goods for coppers?"
As a good citizen, Copernicus warned against speculators, counterfeiters and other swindlers contributing to a decline in the value of money, pointing out that "when good money was in circulation everything was cheap and now the cost of the necessities of life is rising." Anticipating modern economics, the farsighted scholar wanted to see an abundance of all goods, not merely of necessities. He realized that wealth gave the country not only a full stomach, but also conditions favorable to cultural development. Debased money, on the other hand, "causes slothful idleness, neglect of the arts and sciences and a resulting shortage of all goods."
Copernicus had written the memorandum on currency already in 1517–1519, but he delivered it in the form of an address to the Conference of the Estates of Royal Prussia in Grudziądz, held between March 17 and 21, 1522. The next conference of the Estates of Royal Prussia, held on October 3, 1523, also in Grudziądz, debated the monetary reform proposed by Copernicus and resolved to equalize the currencies of Poland and Prussia. The full Latin text of the treatise was published in 1528 under the title **Monetae Cudende Ratio** [The Manner of Minting Money]. This was only three years after the full incorporation of Prussia to Poland. The delegates of Royal Prussia at electoral meetings in 1528 and 1529 were probably familiar with the work of Copernicus on money and perhaps even had it explained by the author himself. The king placed the matter of monetary reform on the agenda of the parliamentary session held in Piotrków, where a commission for currency problems was appointed. It failed, however, to carry out in full the Copernican program of currency reform.

# COPERNICUS DEFENDS WARMIA

Warmia, the homeland of Copernicus, did not enjoy peace in those days. The Teutonic Order, under duke Albrecht, frequently organized raids against Polish territories in Royal Prussia. Irregular bands terrorized villages, robbed merchants on the highway and harassed the population.

Even when such a band was broken up and its leader arrested, as was the case with a certain Hippolytus Wedil, held in the Lidzbark jail, duke Albrecht protested and demanded his release by the Warmia chapter, under threat of reprisals. And Wedil was a savage bandit, who not only robbed the Elbląg merchant Tolkemit, but had both his hands cut off and left him dead on the highway.

The situation in Warmia was deteriorating and firm action was needed to curb the agressiveness of the Teutonic Order.

Nicholas Copernicus took the initiative by addressing to the king of Poland a memorandum on behalf of the Warmia chapter, in which he asked for assistance against persecution by the Teutonic Knights.

His message was an expression of his patriotism, his loyalty to the Polish crown and his hostility toward the Teutonic Order :

"Most Gracious Majesty... We have long wished to complain to you about the injustice we suffer, but we were held back by some timidity and by the majesty of your name, to be invoked in homage rather than with requests. Yet now our situation, and the indignity of recent events, as well as your Majesty's honor compel us to importune your Majesty with our complaints, even though there are more important matters on your mind. It is well known that for seven years now we have been suffering injury at the hands of insolent and criminal men, that we and our subjects have been harassed by fire and sword, by increasingly frequent enemy incursions, that some of us have even been captured, so that we do not have a moment's peace in our homes, exposed to their greed and defenseless. Even the house of God and sacred objects are barely protected from desecration. Finally we are almost prevented from administering justice, while we cannot expect it from others."

Copernicus frankly pointed to those responsible for such attacks.

"Your Majesty is no doubt aware at whose instigation we suffer these atrocities; it is also well known who gives these bandits support, where they are armed and where they find sanctuary with their loot. We have so far endured things with patience, for having taken holy orders we are less expert in the military arts. Yet since the conference at Elbląg had recently resolved that all should take up arms against this plague, and we understood that this was also the intent of your Majesty's edict, we did not shrink from our duty, but became the first to avenge these crimes.
When early this month eight bandits attacked on a public highway one of your Majesty's subjects, a citizen of the city of Elbląg, and cut off both his hands, robbing him of all his possessions, our bailiff quickly called together some of our subjects and pursued these cruel robbers six miles into the territories of the Order, catching up with them before they had time to disperse, as they were engaged in splitting the loot in a marshy forest. He seized one of them a gentleman from the March (the others fled) and brought him here with all the loot the horses and the weapons of the bandits, after having first secured permission from the nobleman to whom that area belongs. The latter now complains that he was wronged, though permission in this case was not even necessary, perhaps in the belief that he might be implicated himself, while the governor at Balga, and even the Grand Master himself demand insistently that the captured bandit together with his loot be delivered to Balga, a matter which our reverend bishop will submit to your Majesty in more detail."

Copernicus then warned against the possible recurrence of such attacks in the near future and clearly blamed the Grand Master of the Teutonic Order.

"That is why the robbers are growing bolder and now molest us with all kinds of provocations, insults and threats. We also note the danger of violence against us by the Grand Master, which we are unable to repel, as our business is praying, not fighting; unless we receive assistance from your gracious Majesty, for which we ask emboldened by the justice of our cause and the danger to our church, which has always had in your Majesty its noblest defender."

In closing the letter, the Warmia chapter offers to the king: "our services and our very lives, ever devoted to your Majesty, our most gracious lord and master."
The letter was dated July 22, 1516, in Warmia.
At about the same time, in the second half of 1516, the Warmia chapter entrusted Copernicus with the administration of the lands under its control. He had under his supervision the towns, villages and population of two-thirds of the diocese of Warmia, including the regions of Pieniężno and Olsztyn. Accordingly he had to move for some time to the castle of Olsztyn which was the administrator's official residence.
Olsztyn castle was built in the form of a massive quadrangle, protected on one side by the river Łyna, which lent great scenic beauty to the setting with its lush vegetation, and

on the other by a deep moat with a drawbridge. Along the walls of the castle were battlements and galleries for the defenders. Olsztyn castle has kept all of its original splendor, as did the market square with its old houses and a medieval town hall. There is still a high, forbidding gate of dark stone in the wall encircling the city, while towers flanked by a church belfry rise on the hill.

To visualize Olsztyn in the times of Copernicus we only have to people these places in our imagination with a crowd in the times of the Middle Ages or the Renaissance. We have to eliminate from our field of vision streetcars and automobiles, filling the roadway with the heavy carts of merchants on their way to northern Warmia.

It was here, at Olsztyn castle, that Copernicus administered the province of Warmia, still harassed by the Teutonic Order. At the conference of the Estates of Prussia in Malbork, in July and then in December 1517, a resolution was adopted calling Royal Prussia to arms. As tension between Poland and the Order was mounting, the papal legate Nicholas Schonberg arrived in Warmia and then proceeded to Królewiec for conciliatory talks with the Grand Master. In the course of these negotiations, in which the king's envoy Jan Łaski also participated, it was suggested that the Teutonic Knights might be moved to Podole, in south-eastern Poland, where they would be charged with fighting the Tartars.

The Grand Master, duke Albrecht, replied to Schonberg's suggestion: "We know very well that the Poles would like to get us to a place which no one can leave or reach." In the meantime the long awaited Polish forces, sent by the king in response to the complaints against the Order, began arriving in Warmia. On December 2, 1518, King Sigismund appeared at the session of the Diet in Toruń escorted by 1,600 mounted men, but this was only the advance guard of a larger army.

Duke Albrecht was also invited to attend the Diet, but when he failed to appear the king proclaimed him an enemy of Poland and accordingly declared war.

The duke responded with subterfuge, as was his habit. He entered Braniewo pretending that he was on his way to a meeting with the king and then seized it. He was planning to capture also nearby Frombork, but a garrison of royal troops protected it against attack.

The war went on with varying fortunes. At first the Polish forces approached Królewiec, but soon afterwards, in August of the same year, the Teutonic Knights besieged Lidzbark — though they failed to take it, owing to the stubborn resistance of the Poles.

As the war continued, the Grand Master threatened Olsztyn itself, after capturing Orneta, twenty-eight miles distant from Olsztyn. It was then that Nicholas Copernicus, re-elected administrator in 1520, directed the defense of Olsztyn against the Teutonic Order. The members of the chapter had fled to Gdańsk, leaving in Warmia only two canons beside Copernicus, John Sculteti and Snellenberg. Copernicus ordered weapons, purchased bombards and organized the defense of the town. He evidently did well, for when the Grand Master approached Olsztyn with his army, he burned some surrounding hamlets, but could not seize the fortified town, either by treachery — as he had done in Braniewo — or by force. He did not even try an assault on the castle, discouraged by the obvious state of readiness of the defenders.

Soon afterward, in April 1521, a four-year armistice was concluded, but intermittent raids by bands instigated by the Order continued. At the regional conferences, for example in Grudziądz, Copernicus and his friend Tideman Giese filed complaints against the Order, charging "damages suffered by the action of the Master."

When the bishop of Warmia died, Copernicus was appointed interim administrator of the diocese, proof of the high esteem in which he was held by the chapter and by the king.

The defense of Warmia against the incursions of the Teutonic Order was finally assured when, on April 10, 1525, the Grand Master knelt before King Sigismund I in the Cracow Market Square, recognizing his suzerainty. At the same time the territories under the control of the Teutonic Order became a secular duchy and the Grand Master its duke, but as a vassal of the crown of Poland. The agreement guaranteed the security of Warmia, which remained an integral part of Poland.

# PRIVATE LIFE

Very little is known about the private life of Copernicus. He lived modestly in his Frombork tower, engaged in scientific work as well as numerous public duties. He did not carry out any ecclesiastic functions, as he probably was never consecrated as a priest. He was rather a lay canon, entrusted by the chapter with various administrative duties. He was chancellor of the chapter, administrator of estates, manager of land settlements. His household was in the care of a distant relative, Anne Szyling from Gdańsk. She was the only woman in Copernicus' entourage throughout his adult life.

One may assume that there was deep affection between them, since she was his only companion. But we would have known nothing about Anne or her relationship with Copernicus if it had not been for the embarrassing interference of his former friend and subsequent superior, John Dantyszek, who was bishop of Warmia during the last years of the astronomer's life.

John Dantyszek, variously appraised by historians, played a somewhat distasteful part in the life of Copernicus. His real name was John Flachsbinder and he was the son of a Gdańsk brewer, a talented, astute and ambitious man. As a young poet he called himself Linodesmos, a name with a Greek flavor, but later he became Dantyszek, a Polish name derived from the Latin form of Gdańsk — Dantiscum.

Smooth and skillful in making friends and gaining people's confidence, witty, rather free in his morals and a typical product of the Renaissance, Dantyszek was obviously destined for a brilliant career. He travelled abroad as envoy of the king of Poland, visiting Italy, the Holy Land and Arab countries; he served as secretary to the king on his journey to Vienna, was ambassador and an outstanding diplomat. He gained the personal favor of emperor Maximilian, from whom he received in 1516 the title of poet laureate, further enhancing his international reputation. Dantyszek's poems were either panegyrics to monarchs, or light and frivolous, even erotic rhymes. As a diplomat he served many masters, which was not unusual at the time; he served as Polish ambassador and then conducted in Venice negotiations on behalf of emperor Maximilian, while on another occasion emperor Charles sent him on a mission to Paris.

He was theoretically in holy orders and had a parish near Cracow, and also derived income from a parish in Gdańsk; he was for a time a canon in Warmia, like Copernicus. All these offices and dignities did not prevent him from having several mistresses in various parts of Europe; he sent them support, primarily for his daughter, Dantyszka. Archbishop John de Weze wrote to his friend Dantyszek: "I visited your Ysopa. I kissed her hands in the Spanish manner, and then her feet in the Roman one... but I kissed not only the hands of the nymph and your lovely daughter, but also their lips..."
Dantyszek, the man of the world, called himself Copernicus' friend for a time. He wrote poems which he proposed as introductions to the works of Copernicus and encouraged him to publish his epoch making treatise.
But as soon as he became his friend's superior, as bishop of Warmia, he turned around completely, apparently forgetting that he had written such lines as:

*I entered the grotto where vice beckoned.*
*In drunken frenzy I possessed them all,*
*I missed none...*

Yet when he sat on the episcopal throne, Dantyszek seemed to suddenly develop austere moral principles, or perhaps he was jealous of the woman devoted to the solitary thinker of Frombork. He knew Anne Szyling as a charming young woman and assumed that she was living with Copernicus as a wife rather than housekeeper, which the former libertine found most reprehensible.
Ostensibly concerned about the morals of the clergy under his authority, Dantyszek poisoned Copernicus' life with a series of letters, asking for the girl's removal from his house, and later even from Frombork.
The first warning concerning Anne was sent by Dantyszek in the second half of 1538, as we know from Copernicus' reply to the bishop's message. He obediently promised that Anne Szyling would leave his house by Easter of 1539. Other letters indicate that she sent her personal effects back to her family in Gdańsk early in 1539.
Copernicus wrote several letters to the bishop, in which he explained the situation, while at the same time trying to postpone his separation from a loyal companion. In a letter dated December 1538 he assured the bishop that he intended to comply with his request but pointed out that it was difficult to find another good housekeeper who was a relative. He promised that the matter would be settled by Easter and pledged all due respect of morals and obedience to the bishop.
Copernicus must have been deeply hurt by Dantyszek's interference, but he carefully kept all bitterness out of his letters. On January 11, 1539, he wrote to the moralistic bishop that he did "all that was requested to do" and hoped that this compliance gave full satisfaction.
Dantyszek, the sanctimonious fraud, was not satisfied by Copernicus' written assurances. Heedless of the alleged friendship with his subordinate, he acted behind his back and gave secret instructions to the vicar of the Frombork parish, Peter Płotowski. The

servile cleric wrote in March of the same year to Dantyszek, reporting about the "Frombork wenches," meaning the housekeepers of Copernicus and canon Sculteti: "One — Alexander's (Sculteti) went into hiding for several days to avoid further interrogation. Doctor Nicholas' woman sent her things to Gdańsk, but she still remains in Frombork."

As to Sculteti, who was persecuted by Dantyszek alongside with Copernicus, he did not yield and eventually married his housekeeper, who bore him a son, Julius. The enmity between him and Dantyszek continued for many years.

Dantyszek hounded the emotional involvement of Copernicus with Anne Szyling with an almost pathological persistence. When he discovered that Copernicus was visiting in Lubawa his loyal friend Tideman Giese, he wrote on July 4, 1539, to Lubawa, charging Copernicus with "entertaining his concubine at secret assignations despite his almost impotent dotage." He urged Tideman Giese to induce Copernicus to "abandon such obscenities and be no longer deluded by Alexander (Sculteti), of whom he says that he alone surpasses in every respect all our brother prelates and canons."

Tideman Giese carried out cautiously and tactfully the request of Dantyszek and reported to him in September a conversation he had with Copernicus. His letter throws a melancholy light on the humiliations and moral anguish endured by the great man at the hands of his former friend and present superior.

Tideman warned Dantyszek not to give too much credence to informers (he meant Peter Płotowski) and reported his conversation with Copernicus as follows:

"He seemed to be much concerned that, even though he promptly complied with Your Excellency's request, some evil minded people still insinuate secret assignations and so on. He denies having seen her since she was discharged; they only spoke briefly when she was on the way to the Królewiec fair. I perceived that he is not suffering as much as some have thought. I ascribe this to his advancing age and continuous studies, as well as the man's virtue and good character; nevertheless I warned him not to create even the appearance of misconduct, and I believe he will be guided by that advice."

Yet another guardian angel appointed by Dantyszek, canon Achacy Trenk of Olsztyn, also reported in September of the same year to the bishop that to he had warned doctor Nicholas Copernicus not to contact his housekeeper and was assured by him "that he would never receive her in his house, nor communicate with her in any manner."

Dantyszek's campaign to deprive Copernicus of the company of the person closest to him, the alleged "wench," finally succeeded. Anne Szyling left Copernicus' home, she lived for some time outside it and then was compelled to leave town. The great man must have suffered deeply, as the letter of Tideman Giese suggests, and his loneliness in the last years of his life no doubt affected adversely both his health and his morale. One can well imagine his anguish as he was forced to continuously apologize to the bishop, at the mercy of vicious gossip and the reports of informers at the service of Dantyszek. All this happened at a time when Rheticus came to Frombork and Copernicus was faced with the dilemma of the publication of his revolutionary work, liable to turn against him not only the bishop, but the entire church, the world of science, in fact all of mankind incapable of accepting a truth reversing all established beliefs.

The loss of a close companion can be a decisive factor if it comes at a crucial point in life, further aggravated by a former friend's treachery. Dantyszek was now aware, of course, that his behavior risked breaking the scientist's spirit at a moment when he had to choose a potentially hazardous course of action, and perhaps depriving the world of one of the great discoveries of all time. The fact that Copernicus, despite the harassment he was already enduring on personal grounds, decided to publish and risk more persecutions testifies to the strength of his character.

# AMONG FRIENDS

No one knows when the world would have become cognizant of the Copernican theory, or whether **De Revolutionibus** would have been published at all, if Georgius Joachimus Rheticus had not entered Copernicus' life.

This highly talented young man, an expert mathematician already at twenty, who later took pride in calling himself a pupil of Copernicus, spread the fame of the Frombork astronomer even before the publication of his great work. It was also he who was responsible for securing a publisher and presenting **De Revolutionibus** to the world.

Rheticus had first heard about the Polish astronomer in that stronghold of Lutheranism, Wittenberg, where some scholars knew about Copernicus. He wanted to meet him in person, prompted both by scientific curiosity and the advice of such Protestant mathematicians and astronomers as John Schoner, Erasmus Reinhold and particularly Philip Melanchthon. The latter, however, later attacked violently the heliocentric theory, as did Martin Luther.

Though associated with these Protestant scholars, Rheticus had no prejudice against Copernicus, nor any reservations about his theory; on the contrary, the young professor from Wittenberg became utterly devoted to Copernicus from the very first. He was modest to the point of self-effacement and always emphasized the greatness of his master. Rheticus had planned a brief visit to Frombork, but it turned into a prolonged stay there. He was captivated by the extraordinary mind of Nicholas Copernicus and by the scope of his ideas, which the astronomer presented to him in tentative form. Forgetting Wittenberg, Rheticus became engrossed in astronomical studies under Copernicus and devoted ten weeks to copying **De Revolutionibus.**

He then hastily wrote his own work, entitled **Narratio Prima de Libris Revolutionum** [First Account of the Books about the Revolutions] in which he outlined the main principles of the Copernican theory and expressed his admiration of the master "that most learned and finest mathematician, the venerable doctor Nicholas of Toruń, canon of Warmia."

This book, written in the form of a letter or report, is full of praise for Copernicus and his work. Rheticus wrote: "This great man, who is now doing me so much good, does not yield to Regiomontanus in any discipline, particularly in astronomy. I would rather compare him to Ptolemy, not that I esteem Regiomontanus to be in any way inferior to him, but because my master had the good fortune, with God's help, to bring his reform of astronomy to its conclusion, while harsh fate took Regiomontanus from among the living before he had time to complete his structure."

Reading these lines we should bear in mind the vast prestige of these names at the time. Comparing Copernicus to them, Rheticus placed him on a pinnacle. He was convinced that he was presenting to the world an astronomer surpassing the world famous Regiomontanus, a genius greater than Ptolemy, who would overthrow his geocentric concept of the universe.

Rheticus, though an enthusiastic supporter of Copernicus, knew the world he lived in and with a view to protecting his master against the attacks of inquisitors and ignoramuses, he stressed that "my teacher desires to follow in the footsteps of Ptolemy, just as Ptolemy followed his elders and those who went before him... My teacher's whole way of thinking is far removed from the idea of abandoning, for the sake of innovation alone, the wisdom transmitted to us by the great philosophers of antiquity."

In summarizing the Copernican heliocentric system, Rheticus wrote that the sun, not earth, is placed in the center of the universe "put there by God as His visible vicar, reflecting the divine majesty of the lord of all creation..."

We find a similar homage to the God-sun in the first volume of **De Revolutionibus**, where — in chapter 10 — Copernicus wrote:

"And in the center of it all sits the sun... some rightly call it the beacon of the world, others its wisdom, or its ruler. Trismegistos called it a visible deity, an all-seeing Electra of Sophocles. Thus the Sun, as if seated on a royal throne, guides the family of planets bustling around it. Earth has at its service the Moon, but as Aristotle observed in his **Of Animals** — the Moon is Earth's closest kin, while the Earth is fertilized every year by the Sun and becomes pregnant to give birth annually."

The Rheticus narrative about Copernicus, summarizing his work quite accurately, was the first public praise received by the great scientist, oddly enough from a pupil with a Lutheran background. The enthusiastic young scholar could hardly have guessed that the harshest condemnation of the Copernican theory was to come from the capital of Protestantism, from its high priests, Martin Luther himself and Philip Melanchthon. In writing his book, Rheticus believed himself to be rendering a valuable service by revealing to the world that in a distant corner of Poland a great, solitary discoverer was initiating the most important intellectual revolution in human history. At the time neither he nor Copernicus himself could visualize the full consequences of that revolution, embracing all aspects of man's spiritual and intellectual life and initiating the Copernican era which still continues.

At the conclusion of his **Narrative** he wrote:

"Let truth prevail! Let the daring be rewarded! Let learning always receive its due respect! Let every master bring forward whatever is useful and present it with moderation, so that it would always be clear that he is striving after the truth alone. My teacher will never be afraid of the judgment of learned and honorable men, but rather will seek it himself."

The **Narrative,** a brief booklet of 38 pages of print, written in the form of a letter to the mathematician John Schoner, aroused wide interest in academic circles. The announcement of a new doctrine of the universe, proclaiming the sun to be still and the earth in orbit, was bound to create a sensation. The first edition of the pamphlet, published in Gdańsk, was sold out immediately and George Winter, a printer in Basel, soon put out a second edition.

The enterprising Rheticus did not confine his efforts to publishing a booklet praising Copernicus and his discovery. He endeavored to publish the entire work of the master and urged Copernicus to prepare a final manuscript for the printers.

**Narratio Prima** was intended to pave the way for the major opus of Copernicus which it accomplished. The Nuremberg printer John Petreius was encouraged by the Rheticus booklet to publish **De Revolutionibus;** he was also urged to do so by Andreas Osiander, who was later to be the book's editor chiefly known for having falsified its introduction.

Copernicus, who had confessed in his letter to Paul III of his hesitation in publishing, now received encouragement from several quarters, among them his trusted friend Tideman Giese, the Capua cardinal Nicholas Schonberg, and — of course — the eager Rheticus. Rheticus came again to Frombork in the middle of 1540 and worked together with him on the manuscripts for publication.

In the summer of 1541 Joachim Rheticus obtained from the university of Wittenberg a leave of absence, so that he could supervise the publication of **De Revolutionibus.** By the middle of 1542 printing was well advanced in the Petreius shop in Nuremberg, as reported in a letter from T. Forsther to J. Schrad in Reuchtlingen: "Prussia gave us a new wonderful astronomer, whose theory we are now printing; the work comprises about a hundred sheets, in which the author asserts and proves that the earth moves and the sky remains immobile. I saw two printed sheets a month ago. A certain Wittenberg master is supervising the proofs."

# A BOOK THAT CHANGED THE WORLD

Let us recall the extraordinary times which served as a setting for the discovery of Copernicus. It was the high noon of the Renaissance; Christopher Columbus had just discovered America, the Spaniard Pinzón reached the coast of Brazil in 1500, and Magellan was soon to circumnavigate the globe, proving by his daring exploit that the earth is indeed a sphere. The life span of Copernicus coincided with those of artists like Michelangelo, Leonardo da Vinci, Wit Stwosz, and scholars like Erasmus of Rotterdam and Callimachus. There was also a dark side. On the papal throne sat a greedy and selfish man given to crime and corruption. It was a period of increasing freedom and exuberance of life, but also of injustice. A contemporary of Copernicus, Savonarola, was hanged and burned, while in Bologna, where our astronomer spent many years. Novara went to the stake because he had uttered some opinions contrary to the Holy Writ or Church dogma.

Copernicus witnessed it all and was well aware of the restraints on human thought, despite the wide intellectual horizons opened by the Renaissance. He could not foresee that some years later his followers and admirers, such as Giordano Bruno or Galileo would suffer torture for merely acknowledging the truth of his discovery.

It was then that Martin Luther nailed to the door of the Wittenberg cathedral his 95 theses, opening a campaign against the pope and the Church which was to win him thousands of followers. Despite his zeal for reform, Luther did not rise above the mental level of other clerics and not only failed to recognize the Copernican discovery, but attacked it in the most primitive and brutal manner.

It was an era that beckoned with the vision of great discoveries and innovations, but at the same time threatened any new ideas likely to undermine the established dogmas or the teaching of the Church fathers. Copernicus continued his studies and his search for truth throughout his life, heedless of the dangers involved.

He finally had a complete work, but did not publish it until the very end of his life, making continuous additions, especially astronomical observations and mathematical calculations. He stated that he had his work ready for scores of years, but he probably meant by this the substance of the discovery rather than the actual content of the book. It contains observations which could not have been made prior to 1535, that is ten years before his death. He wrote in his letter to Pope Paul III, to whom he cautiously dedicated his work, that it had been "kept well hidden not only nine years, but several decades..." Try-

"Nicholas Copernicus in his observatory at Frombork" — painting by Jan Matejko (1873) kept in the Auditorium of the Jagiellonian University in the Collegium Novum.

ing to justify his decision to publish, he admits his previous fears, for it was not until the last years of his life that he was persuaded by other scholars and friends "not to refuse any more to give my work for the common benefit of mankind..."
He obviously feared the reaction of conservative churchmen and scholars, he anticipated the jeers which would greet his statements and that is why he tried to protect his work and himself against attack by dedicating it to the pope.
His dedication read:

"To His Holiness Pope Paul III, Nicholas Copernicus' Preface to the Book on Revolutions."
"I am fully aware, Holy Father, that there are those who, as soon as they hear that I ascribe some movement to earth in my book about the revolutions of heavenly bodies, they will raise a hue and cry against me and my views. I am not so enamored of my own work as to be oblivious of the opinions of others. While I know that the thoughts of a scientist are independent of public opinion — since his purpose is to seek truth, in so far as God permits it to the human mind — I believe that one should steer clear of views clearly contradicting recognized truths. That is why I pondered how absurd people would judge my assertion that the earth moves, which goes counter to the long established conviction that it is immobile and at the center of the universe, and I long hesitated before publishing these books, which I wrote to prove the movement of the earth. I considered following the example of the Pythagoreans and some other thinkers, who were in the habit of transmitting their findings not in writing, but only orally to their closest friends, as witness the letter of Lisius to Hipparchus. They did so, in my opinion, not because they jealously guarded their secrets, as some suspected, but so as not to expose the finest achievements of the long and painstaking labors of great men to the derision and contempt of those who either disdain any study that does not bring them profit, or if they acquire under the influence of others a taste for the noble pursuit of philosophy, are too dull of mind and merely mingle with real scholars like drones among bees. When I considered it all, the fear of ridicule because of the novelty of my theory inclined me to the abandonment of the plans I had for the present work.
But after long delay, and even resistance on my part, I was dissuaded from such a course of action by my friends, foremost among them Nicholas Schonberg, cardinal of Capua, widely known for his universal learning, and my very dear friend, Tideman Giese, bishop of Chełmno, devoted wholeheartedly to theology and all other noble sciences. It was he who pressed me, with many reproaches, to publish this work, which I had kept hidden not only for nine years, but for four decades, and finally let it see the light of day..."

Copernicus was well aware of the revolutionary character of his discovery and of the attacks against the author it might provoke, as he anticipated them in his letter to the pope, seeking his protection:

"To prove both to scientists and laymen that I am not avoiding any criticism, I chose to dedicate this fruit of laborious effort to Your Holiness rather than anyone else, be-

cause even in this distant corner where I live you are held above all others, not only in hierarchic dignity, but also in love of all knowledge, including mathematics. You will therefore be able, through your authority and judgment, to still the slanderous voices, though according to the proverb there is no antidote to the bite of a false accuser.
There will be those who, despite total ignorance of mathematical science, nevertheless claim the right to comment on it, and may condemn and persecute my theory on the basis of some passage in the Bible, wrongly and deviously interpreted to suit their purpose. I care little about them and hold their opinions in contempt as frivolous and inconsequential..."

While writing his letter to the pope, Copernicus had in mind many comments about his discovery, or hypothesis, as it was called; some were favorable, others critical or malicious. Even though **De Revolutionibus** had not yet been published, he was fairly well known as an astronomer.
He owed his reputation partly to the earlier **Commentariolus** in which he had broadly outlined the principles of the structure of the universe, as well as to the **Narratio Prima** of Rheticus.
But the **Commentariolus** was only a fragment, while the major work of Copernicus, the result of several decades of labor, did not appear until the very end of his life.
The first copies were printed by John Petreius in Nuremberg toward the end of March 1543; even if a copy reached Frombork in May 1543, it could not have been consciously seen by the author.
According to a letter from bishop Tideman Giese, a close friend of Copernicus, to Rheticus, Copernicus died on May 24, 1543, of a stroke with paralysis and he had "already many days previously lost his memory and rational mind, while he saw the whole of his work only on the day of his death, as he was breathing his last."
So Copernicus never saw his work in print. He was paralyzed and semi-conscious. The eye which had searched far into space could no longer discern the printed pages. The hand which had held the triangle of the parallactic instrument could not grasp the book holding the secrets of the true movements of the stars and the now mobile earth.
Copernicus missed the joy of seeing his work in print, but he was also spared the distress that the spurious introduction by the editor, Andreas Osiander, would have undoubtedly caused him.
Osiander, a Lutheran theologian who supervised the printing of **De Revolutionibus,** added to the book, without the author's approval, an introduction in which he endeavored to undermine the value of the work, casting doubt upon the validity of the theory advanced by Copernicus; writing for instance "Hypotheses do not have to be true, or even probable, so long as the calculations conform to observable facts... There are other absurdities in this teaching, into which we need not go at this time..."
The unauthorized introduction by Osiander angered Tideman Giese, always loyal to Copernicus. He protested in letters to Joachim Rheticus and the senate of Nuremberg, from which he demanded the punishment of the guilty and the re-printing of the first pages of the book to correct the arbitrary intrusion.

The correction was not made. The introduction, incidentally, failed in its purpose, which was to protect the timorous Lutheran printer against the charge of having published a heretical book. Protestant scholars unleashed a violent attack against Copernicus, no longer living.

What was the actual content of the great work, which by the force of its truth eventually imposed itself on human minds and initiated the Copernican era?

**De Revolutionibus** was not written for laymen, on the contrary, Copernicus himself stressed with pride that "mathematical works are written for mathematicians" and felt that his book could only be appreciated by experts.

While addressing his book to mathematicians, he at the same time chided them in his letter to Pope Paul III for errors and false judgments: "The mathematicians... contradict each other in their investigations. They have so many doubts as to the movements of the sun and the moon that they cannot even calculate a permanent magnitude for the tropical year... Nor have they been able to discover or deduce the most important thing of all namely the structure of the universe and the order of its parts. They acted like an artist who would have painted separately arms, legs, a head and other parts of the body, each quite accurate but from a different model and not in proportion to each other, nor properly joined, thus producing a monstrosity rather than the likeness of a human being. It would seem that in the course of their deduction, or in their method, they either omitted some vital element, or adopted something extraneous and irrelevant."

It was the oddity and the mathematical inaccuracy of the previous astronomical theories that led Copernicus to his discovery. Professor Thomas Kuhn, of the University of California at Berkeley, notes that "for the first time, an astronomer competent in mathematics rejected a tradition of many centuries for scientific reasons, and it was precisely his perception of the inaccuracy of mathematics which started the Copernican revolution."

In his letter to the Pope, Copernicus explained his approach clearly: "Pondering over the inaccuracy of the traditional mathematical sciences in calculating the movements of the spheres of the universe, I was seized by the distressing thought that the philosophers, though otherwise so perceptive in investigating its lesser phenomena, failed to account satisfactorily for the mechanism and the movements of this universe, created for us by the greatest and most perfect of all masters.... This encouraged me to meditate upon the movement of the earth. Although such a thought seemed absurd, I reflected that others before me were free to invent epicycles to their liking in order to explain astronomical phenomena, so there was no reason why I should not be entitled to assume some movement of the earth to give a more satisfactory explanation of the movements of the heavenly spheres.

Assuming the movements which I ascribe to earth, in the present work I realized after many years of observations that if the movements of the other planets are related to the rotation of earth and calculated for the orbit of each separate planet relative to earth, we can not only explain their behavior, but also establish an order with accurate dimensions, binding together all the parts of heaven so that no part of it could be moved without throwing out of balance the other parts and the whole universe."

So much for the dedicatory letter. Let us examine the **De Revolutionibus** itself, described

# THE OPPOSITION

Copernicus was obviously only too well aware of the possible consequences of the publication of his great discovery. But it was not a case of a man concerned over his personal fortunes. To make his discovery he needed not only a brilliant mind and perseverance in his effort, but also the courage that enabled him to discard the thousands of years of tradition behind the geocentric system.

A man with such powers of imaginative vision must have realized that his work would be more than an astronomical and mathematical oddity, that it would undermine the entire philosophical structure of contemporary thought, and perhaps even religious belief.

**De Revolutionibus** carries a tremendous power of conviction, supported by relentless logic and a lucidity of expression typical of genius. It is inspired by a total faith in the truth of the great discovery, even though Copernicus may not have grasped at the time all of its more far-reaching consequences.

He was a recluse, whose only emotional attachment — that to Anne Szyling — was frustrated by the malice of bishop Dantyszek. He could care but little about his personal fate or future, but he might have feared that mankind was not yet ready to accept his work and would greet it with disparagement and ridicule.

Copernicus' discovery encountered violent opposition from the Protestants, while it was not attacked at first by the Catholic Church. A book written by a canon and dedicated to the pope did not appear heretical, especially as the Church fathers were not fully cognizant of the long-range implications of the Copernican revolution.

The Protestants, however, attacked Copernicus bitterly, led by Luther himself. One of the first to criticize him was Philip Melanchthon, a leading Lutheran theologian, who wrote: "Some people think it commendable and clever to elaborate something as absurd as the Sarmatian astronomer who stops the sun and sets the earth on the move. Wise rulers should restrict frivolous talent."

He dealt with the subject again, some years after Copernicus' death, in a book entitled **Initia Doctrinae Physicae** [Introduction for Physical Science 1549].

"Our eyes attest to the fact that the sky turns every twenty-four hours. Yet some, whether in search of novelty, or to display their talent for speculation, ponder the movement of the earth... Such facetious ideas had been tried before. Some masters try many ideas just to exercise their minds, but to insist on plainly absurd assertions is dishonest and sets a bad example. Even though some may laugh at a physicist invoking divine testimony, we believe that philosophy should rely on heavenly verdicts. The psalmist clearly speaks about the sun moving... Another psalm, in referring to the earth, says 'it is firm in its foundations and will not move for ever and ever.' In the first chapter of the Ecclesiastes one reads: 'Let the earth be forever immobile, and let the sun rise and set.' Reassured by this divine testimony we hold on to the truth and will not be deceived by those who think that bringing confusion into science is a glorious achievement."

But Martin Luther made a more vicious comment on the Copernican discovery. Laughable though the Church reformer's sarcastic remarks may sound today, they may have dissuaded many potential supporters of Copernicus at the time. "There is talk of a new astrologer, who wanted to prove that the earth rotates, rather than the firmament with the sun and the moon; just as if someone was sitting in a carriage or a boat in motion, thinking himself immobile and the ground and the trees moving. But this is the fashion today, anyone who wants to pass for a wise man has to think up some novelty and tout it as the best. That fool wants to overturn the whole science of astronomy! But, as the Bible tells us, Joshua told the sun to stop, not the earth!"

The Copernican theory was attacked not only by Protestant astronomers or scientists, but also by writers and thinkers. For example the sixteenth century French author Du Bartas, in his long cosmological poem **The Week, or the Creation of the World**, derided:

*The clerks who think*
*That neither heavens nor stars do turn at all,*
*Nor dance about this great round earthly ball;*
*But th'earth itself, this massy globe of ours,*
*Turns round-about once every twice-twelve hours;*
*And we resemble land bred novices*
*New brought aboard to venture on the seas;*
*Who, at first launching from the shore, suppose*
*The ship stands still, and that the ground it goes...*
*So never should an arrow, shot upright,*
*In the same place upon the shooter light;*
*But would do rather as at sea a stone*
*Aboard a ship upward uprightly thrown;*
*Which not within board falls, but in the flood*
*Astern the ship, if so the wind be good.*
*So should the fowls that take their nimble flight*
*From western marches towards morning's light...*
*And bullets thundered from the cannon's throat*
*(Whose roaring drowns the heavenly thunder's note)*
*Should seem recoiled; since the quick career*
*That our round earth should daily gallop here,*
*Must needs exceed a hundredfold, for swift*
*Birds, bullets, winds; their force, their drift.*
*Arm'd with these reasons, 'twere superfluous*
*T'assail the reasons of Copernicus;*
*Who to save better of the stars th'appearance*
*Unto the earth a threefold motion warrants.*

Another eminent French writer of the XVI-th century, Jean Bodin, wrote: "No one in his right mind, no one with a minimum knowledge of physics could believe that the earth, heavy and inert because of its vast mass, rotates around its own axis and revolves around the sun. Even the slightest tremor of the earth would cause houses and fortresses, cities and mountains to crumble down."

Calvin joined the chorus of anti-Copernican attacks and wrote in his **Commentary to the Book of Genesis**: "The Lord fixed the circumference of the world so as to make it immobile. Who would dare place the authority of Copernicus above that of the Holy Writ?"

William Gnapheus of Elbląg launched a satirical attack against Copernicus in his lifetime, by writing a comedy entitled **Morosophus** (The Foolish Scholar) in which he clearly ridiculed Copernicus, though never mentioning his name. The intent of the dialogue, involving a "pompous philosopher" and "comical astrologastrus" was quite evident. The play was produced at the school theatre of Elbląg and published in 1541; it is notable that Gnapheus dedicated his work to duke Albrecht, the former Grand Master.

Some of Copernicus' critics not only questioned the validity of his conclusions, and treated the heliocentric idea jocularly, but urged brutal reprisals against the astronomer. For example, Francesco Maurolyco wrote: "Some even praise Nicholas Copernicus, who deserves not censure but a whipping," while Scagliero raged: "Such writings should be expunged, or their authors flogged."

Despite these outbursts, the Copernican theory was at first treated by the Catholic Church as a hypothetical concept, in accordance with the presentation by Osiander. Several decades later, however, the potential danger of the Copernican doctrine to the whole established order of the world became apparent.

It was increasingly realized that if the earth as one of the planets indeed moves many passages of the Bible would have to receive a new interpretation. People could well ask where the divine throne, heaven and hell were located. The division between the kingdom of heaven and the vale of tears was shaken, since the earth was no longer the center of the universe. There might be other worlds with other creatures. How would one understand then the creation of the world with Adam and Eve, made in the likeness of God?

**De Revolutionibus** moved not only the earth, but also thoroughly shook the entire structure of medieval philosophy. Toward the end of the XVI-th century the scholars of the Church realized the dangers of the Copernican doctrine, but they were unable to adapt the teaching of the Church to it, nor did they even try to reconcile the heliocentric concept with religious dogma.

Instead, they initiated a campaign against the Copernican theory and its growing supporters. The Holy Office of the Inquisition began to operate and in 1600, several decades after the death of Nicholas Copernicus, the philosopher Giordano Bruno went to the stake, for enthusiastically advocating the Copernican theory, as well as other' heresies.' He became the symbol of fearless dedication to truth and intellectual freedom at any price.

In 1616 another great champion of Copernicus, Galileo Galilei, was charged for the first time by the Inquisition and let off with a warning; however, he was summoned again in 1633 for propagating Copernican ideas.

Even today, centuries later, the verdict of the Holy Office against Galileo, and indirectly

against Copernicus as well, makes harrowing reading, as it recalls a drama symbolic of the eternal conflict between authority and truth.
We see the astronomer kneeling before the commissioners of the Inquisition, charged not only with believing the Copernican theory, but even with developing it in his **Dialogue on Two Principal World Systems: the Ptolemaic and the Copernican.**
We see the fine features of Galileo, circled with a greying beard, his wise eyes — so recently filled with joy for having confirmed by means of a telescope the findings of his great predecessor — now clouded by anguish at the verdict of the Holy Office.
The seventy year old scientist had to listen, like a criminal and heretic, to words addressed to him and Copernicus alike:

"Whereas you, Galileo, son of the late Vincenzo Galileo, Florentine, aged seventy, were in the year 1616 denounced to this Holy Office for holding as true the false doctrine taught by some that the Sun is the center of the world, and immovable, and that the Earth moves, and also with a diurnal motion; for having disciples to whom you taught the same doctrine; for holding correspondence with certain mathematicians of Germany concerning the same; for printing a letter entitled **On the Sunspots** wherein you developed the same doctrine as true; and for replying to the objections from the Holy Scriptures which from time to time were urged against it, by glossing the same Scriptures according to your own meaning; and whereas there was produced a copy of a document in the form of a letter, purported to be written by you to one formerly your disciple, and in this diverse propositions are set forth, following the position of Copernicus, which are contrary to the true sense and authority of the Holy Scripture.
The Holy Tribunal being therefore of intention to proceed against the disorder and mischief thence resulting, which went on increasing to the prejudice of the Holy Faith, by command of His Holiness and of the most eminent Lord Cardinals of this supreme and universal Inquisition, the two propositions of the stability of the Sun and the motion of the Earth were set by the Theological Counsels as follows: The proposition that the Sun is the center of the world and does not move from its place is absurd, false philosophically and formally heretical, because it is expressly contrary to Holy Scripture.
The proposition that the Earth is not the center of the world and immovable, but that it moves, and also with a diurnal motion, is equally absurd and false philosophically, and theologically considered at least erroneous in faith.
But whereas it was desired at that time to deal leniently with you, it was decreed at the Holy Congregation held before His Holiness on February 25, 1616, that His Eminence the Lord Cardinal Bellarmine should order you to abandon altogether the said false doctrine and, in the event of your refusal, that an injunction be imposed on you by the Commissioner of the Holy Office to give up the doctrine and not to teach it to others, and failing your acquiescence in this injunction that you should be imprisoned. And in execution of this decree, on the following day, at the Palace, and in the presence of His Eminence the said Cardinal Bellarmine, after being gently admonished by the said Lord Cardinal, the command was enjoined upon you by the Father Commissioner of the Holy Office of that time, before a notary and witnesses, that you were altogether to

abandon the false opinion, and in future not to hold or defend or teach it in any way whatsoever, neither verbally nor in writing, and upon your promising to obey, you were dismissed...

But whereas a book appeared last year in Florence, the title of which shows that you were the author, this title being **Dialogue of Galileo Galilei on the Great World Systems;** and whereas the Holy Congregation was afterwards informed that through the publication of the said book, the false opinion of the motion of the Earth and the stability of the Sun was daily gaining ground, the said book was taken into careful consideration, and in it there was discovered a potent violation of the aforesaid injunction that had been imposed upon you, for in this book you have defended the said opinion previously condemned and to your face declared to be so, although in the said book you strive by various devices to produce the impression that you leave it undecided, and in express terms as probable: which, however, is a most grevious error, as an opinion can in no way be probable which has been declared and defined to be contrary to divine Scripture.

Therefore, having seen and maturely considered the merits of your cause, together with your confessions and excuses above-mentioned, and all that ought justly so be seen and considered, we have arrived at the underwritten final sentence against you:

Invoking, therefore, the most holy name of our Lord Jesus Christ and of His most glorious Mother, ever Virgin Mary, by this our final sentence, which sitting in judgment, with the counsel and advice of the Reverend Masters of sacred theology, and Doctors of both Laws, our assessors, we deliver in these writings... we say, pronounce, and declare that you, the said Galileo, by reason of the matters adduced in trial, and by you confessed as above, have rendered yourself in the judgment of this Holy Office vehemently suspected of having believed and held the doctrine — which is false and contrary to the sacred and divine Scriptures — that the sun is the center of the world and does not move from east to west, and that the Earth moves and is not the center of the world; and that an opinion may be held and defended as probable after it has been declared and defined to be contrary to the Holy Scripture; and that consequently you have incurred all the censures and penalties imposed and promulgated in the sacred canons and other constitutions, general and particular against such delinquents. From which we are content that you be absolved, provided that first, with a sincere heart and unfeigned faith, you abjure, curse, and detest before us the aforesaid errors and heresies and every other error and heresy contrary to the Catholic and Apostolic Roman Church in the form to be prescribed.

And in order that this your grave and pernicious error and transgression may not remain altogether unpunished and that you may be more cautious in the future and an example to others that they may abstain similar delinquencies, we ordain that the book of the **Dialogue of Galileo Galilei** be prohibited by public edict.

We condemn you to the formal prison of this Holy Office during our pleasure, and by way of salutary penance, we enjoin that for three years to come, you repeat once a week the seven penitential Psalms."

Tortured by the Inquisition and morally shattered, Galileo now had to utter the oath of. abjuration of the teachings of Copernicus, to abjure the truth and declare that the judgement of ignorance was right. Galileo thus knelt and declared:

I, Galileo, son of the late Vincenzo Galileo, Florentine, aged seventy, arraigned personally before this tribunal and kneeling before you, most Eminent and Reverend Lord Cardinals Inquisitors-General against heretical depravity throughout the entire Christian commonwealth, having before my eyes and touching with my hands the Holy Gospels, swear that I have always believed, do believe and by God's help will in the future believe all that is held, preached and taught by the Holy Catholic and Apostolic church. But, whereas, after an injunction had been judicially intimated to me by this Holy Office to the effect that I must altogether abandon the false opinion that the Sun is the center of the world and immovable, and that the Earth is not the center of the world, and moves, and that I must not hold, defend, or teach in any way whatsoever, verbally or in writing, the said false doctrine, and after it had been notified to me that said doctrine was contrary to Holy Scripture, I wrote and printed a book in which I discuss this new doctrine already condemned, and adduce arguments cogent in its favor without presenting any solution of these, I have been pronounced by the Holy Office to be vehemently suspected of heresy, that is to say of having held and believed that the Sun is the center of the world and immovable, and that the earth is not the center, and moves.... Therefore desiring to remove from the minds of Your Eminences, and all faithful Christians, this vehement suspicion justly conceived against me, with sincere heart and unfeigned faith I abjure, curse and detest the aforesaid errors and heresies and generally every other error, heresy, and sect whatsoever, contrary to the Holy Church, and I swear that in future I will never again say or assert, verbally or in writing, anything that might furnish occasion for a similar suspicion regarding me; but should I know any heretic, or person suspected of heresy, I will denounce him to this Holy Office or to the Inquisitor or Ordinary of the place where I may be. And in the event of my contravening (which God forbid) any of these my promises and oaths, I submit myself to all the pains and penalties imposed and promulgated in the sacred canons and other constitutions, general and particular, against such delinquents. So help me God and these his Holy Gospels, which I touch with my hands....

I, the said Galileo Galilei, have abjured, sworn, promised and bound myself as above; and in witness of the truth thereof I have with my hand subscribed to the present document of my abjuration and recited it word for word at Rome, in the convent of the Minerva, this twenty-second day of June, 1633. I, Galileo Galilei, have abjured the above with my own hand."

This abjuration of Copernicus' discovery occurred 90 years after the publication of **De Revolutionibus** and lasted for centuries. Not until 1822 was Copernicus' book taken off the **Index** of the Vatican, and his discovery admitted to be true.

# COPERNICUS TO ARMSTRONG

The Copernican theory opened a new era. The collapse of the geocentric system shook the ideological structure of the world, shattering the foundations of human thought, which were unchallenged through centuries.

Copernicus based his work on the astronomical tradition available to him, but in doing so he demolished that tradition, creating a new vision of the universe.

The very concept of the universe underwent a profound change, triggering a wave of revolutionary innovation in all branches of science, social thought, ideology and creativity. The ideas of Copernicus, developed by his successors, such as Galileo, Kepler, and Newton, collided with the force of an avalanche with all the established doctrines and hitherto sacrosanct dogmas. Mankind entered the age of science, invention and bold investigation of the laws of nature.

The Copernican era opened a new perspective of the universe. The earth became a small planet, but also a vantage point for observing the vast spaces of the universe. Copernicus reduced the earth to the size of a small point relative to the 'mass of the world,' but he increased at the same time the human mind's capacity to probe the mysteries of the cosmos and those of earth itself. The hermit of the Frombork tower discovered "the amazing harmony of the world" and set in motion, not only the earth, but also man himself, launching him into orbit in the vastness of space.

Nicholas Copernicus could hardly have foreseen that several centuries after his death the first man would fly an artificial star around the earth, and soon afterwards another would set foot on the moon, to see their planet earth circling the immobile sun. These men, Gagarin and Armstrong, explored space along the paths discovered by Copernicus and one of them, upon his return from the moon, said that he owed his achievement to Copernicus, Galileo and Newton.

The Copernican era is still on.

Nicholas Copernicus — copperplate by Jeremy Falco c. 1645.

# TORUŃ

Toruń. Nicholas Copernicus — part of a statue unveiled in 1853, sculpted by Frederic Trieck.

Toruń. View of the roofs and gables of the Old Town houses.

Toruń. East view of the Gothic Town Hall — 14th century (top floor dating from 1603).

Toruń. Statue of Nicholas Copernicus at the south-east corner of the Old Town Hall. ▶

◀ Part of the 13th century cellar of the Old Town Hall used for storing wine and beer in the Middle Ages.

◀ Toruń. Part of the ground floor of the Old Town Hall, which houses an exhibition of medieval art.

Toruń. Old Town Hall — part of the Great Hall, also known as the Burghers' Hall with a beamed ceiling decorated in polychrome between 1730 and 1740, and a gallery with portraits of Toruń's mayors and councillors.

Nicholas Copernicus Street in Toruń (former St. Anne Street) with Copernicus' House built in the second half of 15th century, and restored in 1960—1963. ▶

Toruń. Napoleon before what is thought to be Copernicus' House, after 19th century litograph.

Toruń. Bottom part of the façade of Copernicus' House.

Toruń. Upper part of the façade of Copernicus' House (copestone reconstructed).

Toruń. View of the Old Town with the massive silhouette of Gothic church of St. John dating from the 14th and 15th centuries.

Toruń. Commemorating plaque to Nicholas Copernicus donated by Toruń physician Melchior Pirnesins around 1580.

Toruń. Gothic baptismal font from the second half of 13th century in the church of St. John.

◀ Toruń. Part of the 14th century town walls close to the Burghers' Court, seen from the River Vistula.

Toruń. Bridge Gate, originally called Ferryboat Gate, built in 1432 by master Jan Gotland (copestone dating from 19th century).

Toruń. Gdanisko ("Dansker") — part of the sanitary facilities in the Teutonic Knights' castle from the beginning of 14th century.

# CRACOW

Cracow. Wawel Royal Castle — view of the north-east side, called the Kurza Stopka, 15th century. ▶

Cracow. View of the city, after a woodcut in Hartmann Schedel's "World Chronicle", 1493.

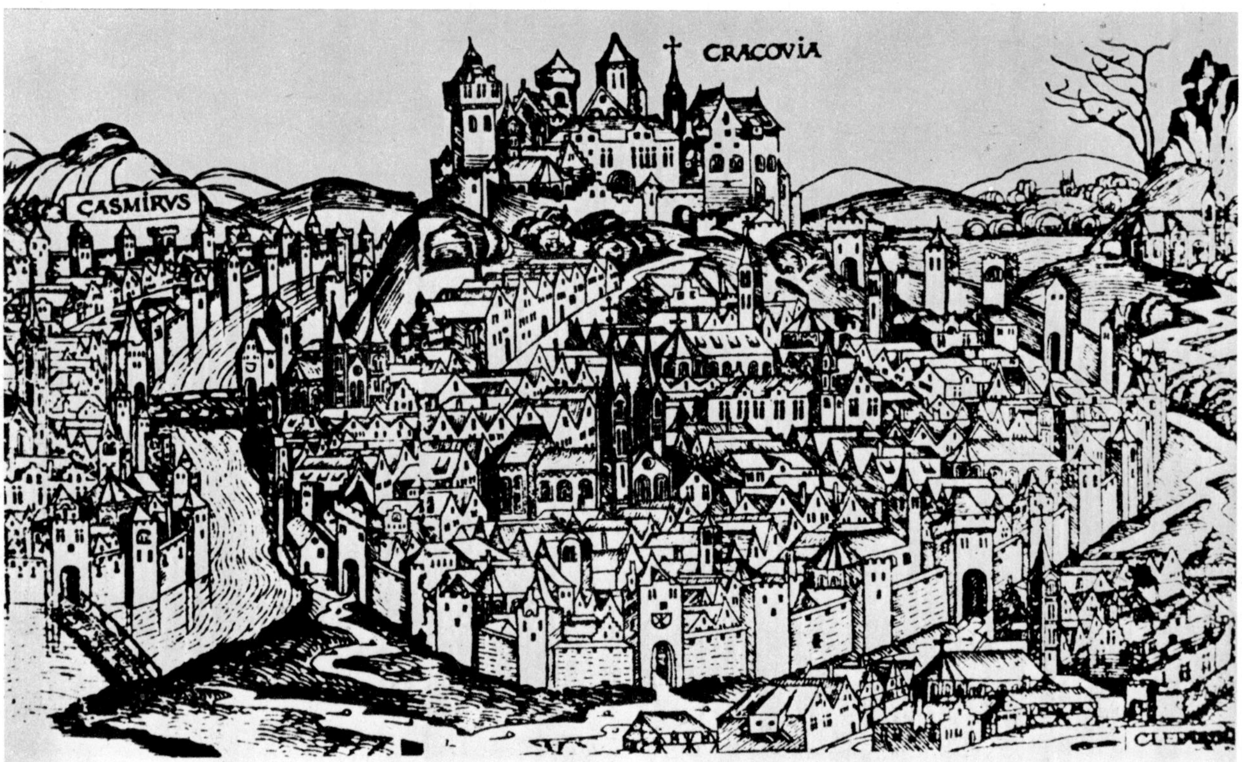

◀ Wawel Hill with a view of the south wall of the Cathedral, built in the 14th and 15th centuries, as well as the chapels of the Potockis, Vasas and Sigismund (1517—1533).

Cracow. Renaissance courtyard of Wawel Castle, dating from 1507—1536.

◀ Cracow. Tomb of the King Casimir Jagiellon in Wawel Cathedral; executed by Wit Stwosz in 1492.

Cracow. Part of the tomb of Casimir Jagiellon.

Cracow. Sarcophagus of Queen Jadwiga (founder of the Cracow Academy) in Wawel Cathedral; executed by Antoni Madeyski in 1902.

◀ Cracow. East wall of the 15tn century Collegium Maius (completely reconstructed in 1947—1963) — the oldest building of the Jagiellonian University.

Cracow. A lecture at the Cracow Academy — after 15th century woodcut.

Cracow. The Founding of the Cracow Academy — after 15th century woodcut.

200. HERBY

Sprawy święte á pámięci wieczney godne thego Królá Władysławá Jágełá Wielkiego X. Litewskiego/ á naprzod záłożenie Akádemiey Krákowskiey: skąd káżdey cno ty dobrey początek wychodźi/ Roku 1401.

Ly masz mátkę cnot świętych od czásu dawnego/
Wárownie záłożona przez Królá sławnego/
Przypátrzie sie iey synom ktore wychowáłá /

Cracow. The arcaded courtyard of the Collegium Maius.

Cracow. Stairway in the cloister of the Collegium Maius.

Cracow. Baroque portal leading into the courtyard of the Collegium Maius.

Cracow. The Common Hall (*Stuba Communis*) — one of the oldest halls in the Collegium Maius, rebuilt at the turn of the 17th century.

Cracow. The Common Hall.

Cracow. Portrait of Nicholas Copernicus' father — copy dating from the 17th century, on display at the Collegium Maius. ▶

Cracow. Martin Bylica's globe of the sky with mounted astrolabe dating from 1480—1490, on display at the Collegium Maius.

NICOLAVS COPERNICVS PATERNICO
NATI 1473 19 FEBRVAPY
NNES BROSCIVS CVRZELOVIENS DEPINGI CVRAVIT

Cracow. Arabian astrolabe dating from 1054, on display at the Collegium Maius.

Cracow. Clock with armillary sphere, called the Jagiellonian Clock, dating from the first quarter of the 16th century, on display at the Collegium Maius.

Circulus arcticus
Polus zodiaci
Tropicus
Cancri
Zodiacus
Equinoctialis
Tropicus
Capricorni
Polus
mundi
Antarticus
PTOLE MEVS
IOHANES DE MONTE

Claudius Ptolemy (second century AD) author of the "Almagest," the work in which he expounded his geocentric theory of the universe.

Document confirming Copernicus' enrolment in the Cracow Academy in 1491.

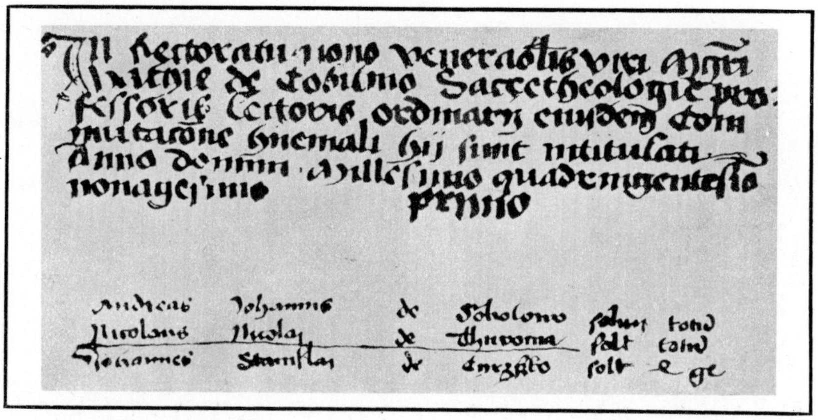
In rectoratu novo venerabilis viri magistri
Mathie de Kobilino Sacre theologie pro-
fessoris lectoris ordinarii eiusdem com-
mutatione hiemali hii sunt intitulati
anno domini Millesimo quadringentesimo
nonagesimo
primo

| | | | | | |
|---|---|---|---|---|---|
| Andreas | Johannis | de | Sokolowo | solvit | totum |
| Nicolaus | Nicolai | de | Thorunia | solvit | totum |
| Johannes | Stanislai | de | Krzyszkowo | solvit | 2 gr |

◀ Title page of Ptolemy's "Almagest."

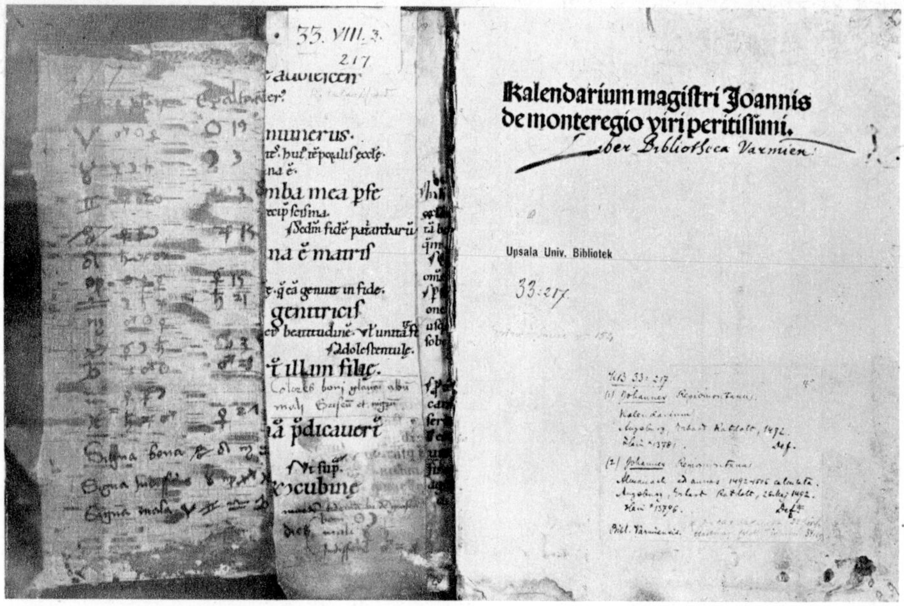


Title page of the "Astronomical Tables" of the German astronomer and mathematician Jan Müller — Regiomontan (1436—1476) — from the library of Nicholas Copernicus with his annotations.

A page from Regiomontan's "Astronomical Tables" with Copernicus' annotation.

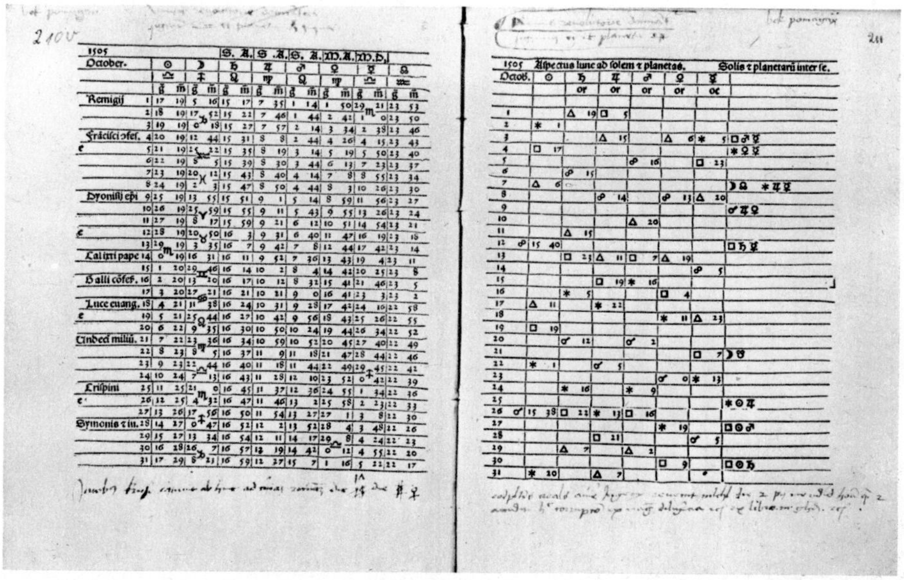

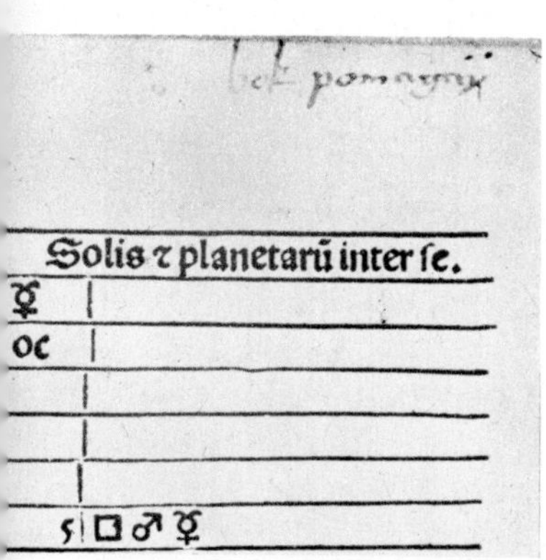


The top part of a page from the "Astronomical Tables" of Regiomontan showing Copernicus' annotations.

◀ Cracow. Statue of Nicholas Copernicus executed by Cyprian Godebski, unveiled in the courtyard of the Collegium Maius in 1900, now standing near the Collegium Novum.

Astronomer and Mathematician Wojciech of Brudzewo (1445—1497), professor of the Cracow Academy during Copernicus' student days in Cracow.

Cracow. Book containing the manuscript of Copernicus "De Revolutionibus..." — kept in Jagiellonian Library.

Nicholas Copernicus—portrait from second half of 16th century, placed on astronomical clock in Strassbourg Cathedral.

Cracow. Manuscript of Copernicus "De Revolutionibus..." open at the drawing of the heliocentric system.

Cracow. Part of the Market Place with the view of the late 14th century Gothic Cloth Hall reconstructed in 1557—1559 by Jan Maria Padovano (arcades and projection dating from 19th century).

Cracow. Town Hall Tower in the Market Place, surmounted by the Baroque cupola.

Cracow. Church of St. Adalbert, built in 11th century and completely reconstructed in the first half of the 17th and in the 18th centuries.

LOVE TO PRAY —
feel often during
the day the need
for prayer
and take trouble
to pray.
Prayer enlarges the
heart until it is
capable of
containing God's
gift of Himself.
Ask and seek, and
your heart will
grow big enough
to receive Him and
keep Him as your
own.

◀ Cracow. Part of the Old Town with the view of the façade of Virgin Mary Church, dating from the 14th and 15th centuries.

Cracow. Part of the main scene — the Dormition of the Virgin, of the polyptych in Virgin Mary Church.

Cracow. Late-Gothic polyptych in Virgin Mary Church, executed by Wit Stwosz in 1477—1489.

Cracow. The Florian Gate, built in 1307 — the main entrance to the City in the Middle Ages. ▶

Cracow. The Barbicane, built to reinforce the Florian Gate in 1489—1499.

Cracow. Part of Grodzka Street with a view of Romanesque towers of the church of St. Andrew built in the 12th century and surmounted by a Baroque cupola.

Cracow. Part of the Baroque portal of the church of St. Barbara with the coats-of-arms of the Crown and of Lithuania.

# WROCŁAW

Wrocław. Part of a perspective, after 17th century copperplate.

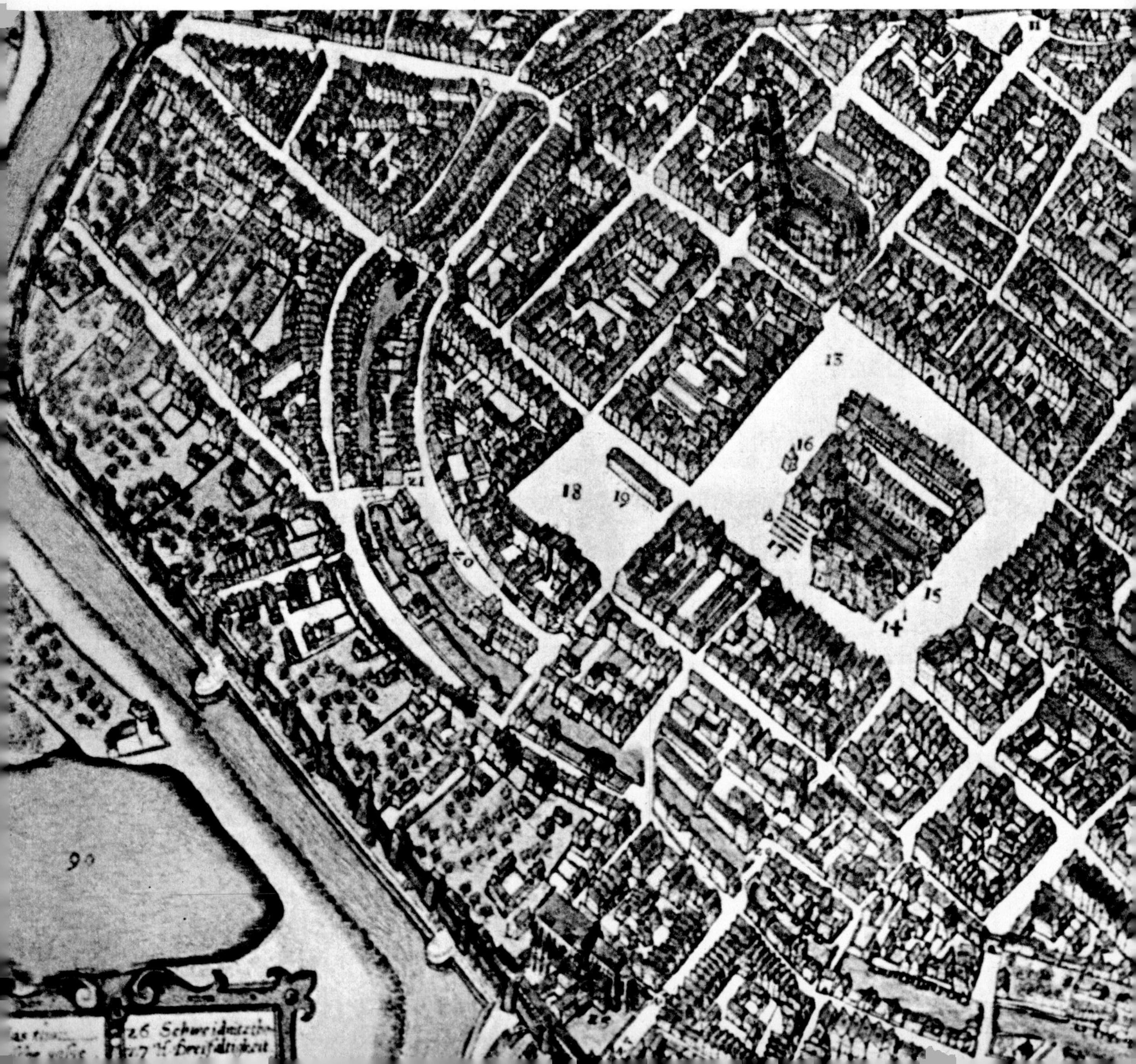

Wrocław. Panorama of the Old Town.

Wrocław. Solny Square showing the frontages of the houses on the south side.

Wrocław. The Gothic canonry of the church of St. Giles. ▶

Wrocław. View of Ostrów Tumski from the River Odra. On the left the church of St. Martin, built in the 13th century, reconstructed in the 16th century; rebuilt after the Second World War. In the distance the Church of Holy Cross built in 13th and 14th centuries.

◀ Wrocław. The late-Gothic 14th century Town Hall, reconstructed in the 15th and 16th centuries.

Wrocław. Part of an exhibit in the Aldermen's Chamber in the Town Hall.

# BOLOGNA

Bologna. View of the town after a woodcut in Hartmann Schedel's "World Chronicle," 1493.

Bologna. View of the town from the top of the Torre Asinelli.

◀ Bologna. "Due Torri" — the leaning towers of Asinelli and Garisenda built in the first half of the 12th century.

Bologna. Piazza Maggiore — the medieval town centre, showing parts of 13th/15th century Town Hall and the palace of King Enzo, dating from the same period.

Bologna. Palazzo del Podestà, dating from the 13th-15th century with the Torre dell'Arengo.

Bologna. Characteristic colonnades on the Via Zamboni in the Old University district. ▶

Bologna. The rooftops of the Old University quarter. ▶

Bologna. Square in the Old University district.

Bologna. Courtyard of the cloister attached to the church of San Stefano (11th-13th centuries).

Bologna. Statue of Nicholas Copernicus in the Hall of the University. Bust carved by the Polish Sculptress, Jadwiga Bohdanowicz.

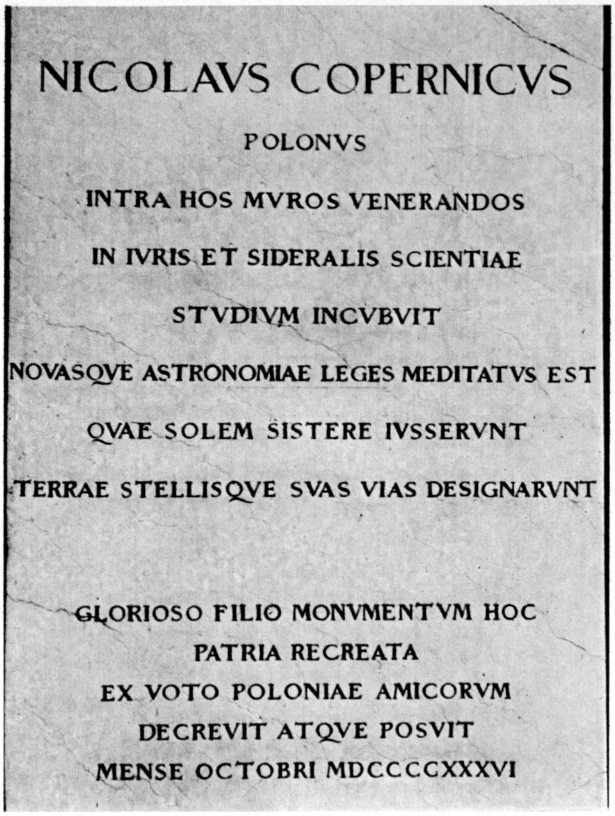


Bologna. Inscription on the plinth of the statue.

Bologna. Part of the 15th century tombstone of Antonio Bentivoglio in the church of San Giacomo Maggiore — university students attending a lecture.

Bologna. Young people under the colonnades of the university.

Bologna. The Neptune Fountain, dating from 1564, the work of Jan de Douda called Gianbologna.

Christopher Columbus (1451—1506).

Girolamo Savonarola (1452—1498), Dominican theologian and reformer.

Bologna. View of the façade of the Basilica San Petronio (patron of the town) during restoration; the basilica built in 1383—1659, was designed by Antonio di Vincenzo.

# ROME

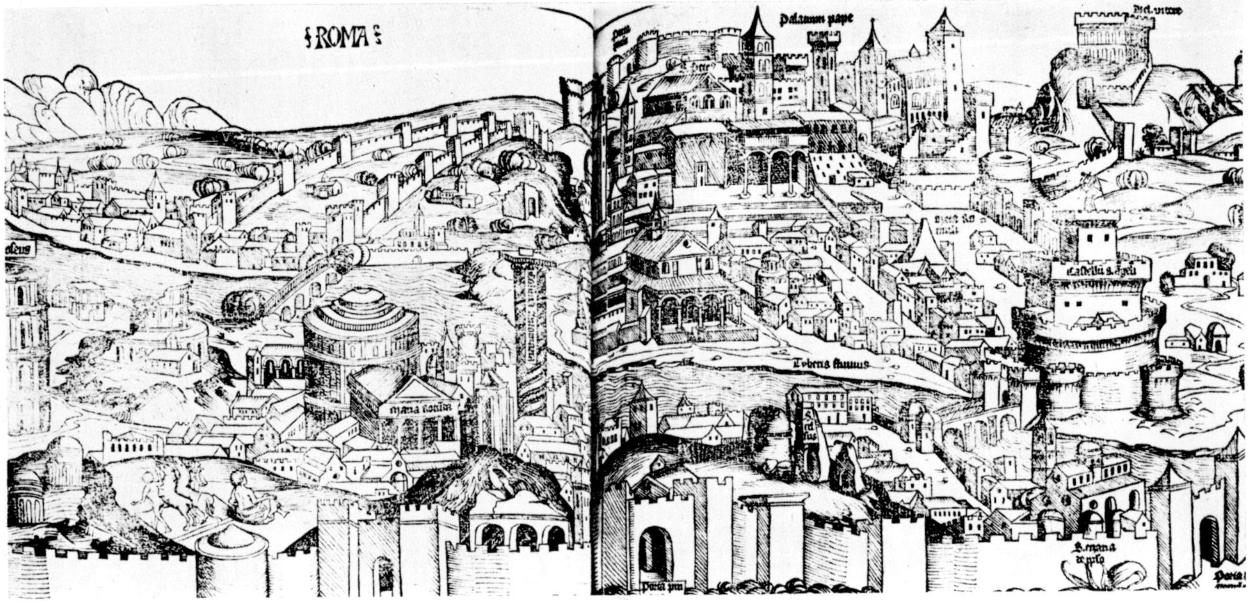

Rome. View of the city, after a woodcut in Hartmann Schedel's "World Chronicle," 1493.

A lecture by Nicholas Copernicus in Rome in 1500, etched Henryk Redlich in 1876 after a drawing by Wojciech Gerson.

Rome. Showcase with a collection of medals commemorating Nicholas Copernicus in the museum of the great astronomer set up by Artur Wołyński at Astronomical Observatory in 1876.

Rome. The Via Appia, the oldest paved Roman road, called the "King of roads," built in 312 in the time of Appius Claudius.

Rome. Part of Palazzo Venezia, built in the 15th century.

◀

Rome. The castle of San Angelo originally the mausoleum of Emperor Hadrian, built in the years 135—139, loggia facing Tiber, built during the Renaissance.

Rome. The Coliseum, built in the years 72—80 under Emperor Titus. ▶

Rome. View of the church Santa Maria di Loreto, built by A. Sangallo in 1507, as seen from the Trajan Forum. On the left the Trajan Column. ▶

# PADUA

Padua. Barrows in front of Palazzo della Ragione. ▶

Padua. View of the city after a woodcut in Hartmann Schedel's "World Chronicle," 1493.

Padua. Medieval coats-of-arms on the walls of the Palazzo della Ragione. ▶

Padua. Palazzo della Ragione built in the 12th, 14th and 16th centuries. View from the south.

PIETRO COZZO
PADOVA REPUBLICA

Padua. Piazza dei Signori with the view of the 15th/16th century Palazzo del Capitanio.

Padua. Sun-dial with signs of Zodiac on the tower of the 16th century Town Hall.

Padua. Frontal of the Basilica of San Antonio, dating from the 13th to 15th centuries.

Padua. "Theatrum Anatomicum" — prosectorium in the Old University, built in 1594 by Fabrizio d'Acquapendete (the first in Europe).

AULA MAGNA

◀ Entrance to the Great Hall of the Old University of Padua.

Padua. Cloisters in the Old University.

# FERRARA

Ferrara. View of the city after a woodcut in Hartmann Schedel's "World Chronicle," 1493.

Ferrara. Map of the city showing the fortifications, after a 17th century copperplate.

Ferrara. 12th-14th century Romanesque Cathedral with its elaborate Gothic façade.

Ferrara. Lions in front of the Cathedral.

IL GIORNO
IL GIORNO

◀ Ferrara. Entrance to the 13th century Palazzo del Municipio, expanded in the 15th century with the addition of statues carved by Pietro Benvenuti.

Ferrara. Castle of the Princes d'Este, dating from the 14th century, reconstructed in the 16th century.

Ferrara. The castle towers.

Ferrara. Inner defence moats of the castle. ▶

Ferrara. Official records with an entry noting the conferment of a Doctor's Degree on Nicholas Copernicus.

Ferrara. Page with the text of the conferment of a Doctor's Degree on Nicholas Copernicus. ▶

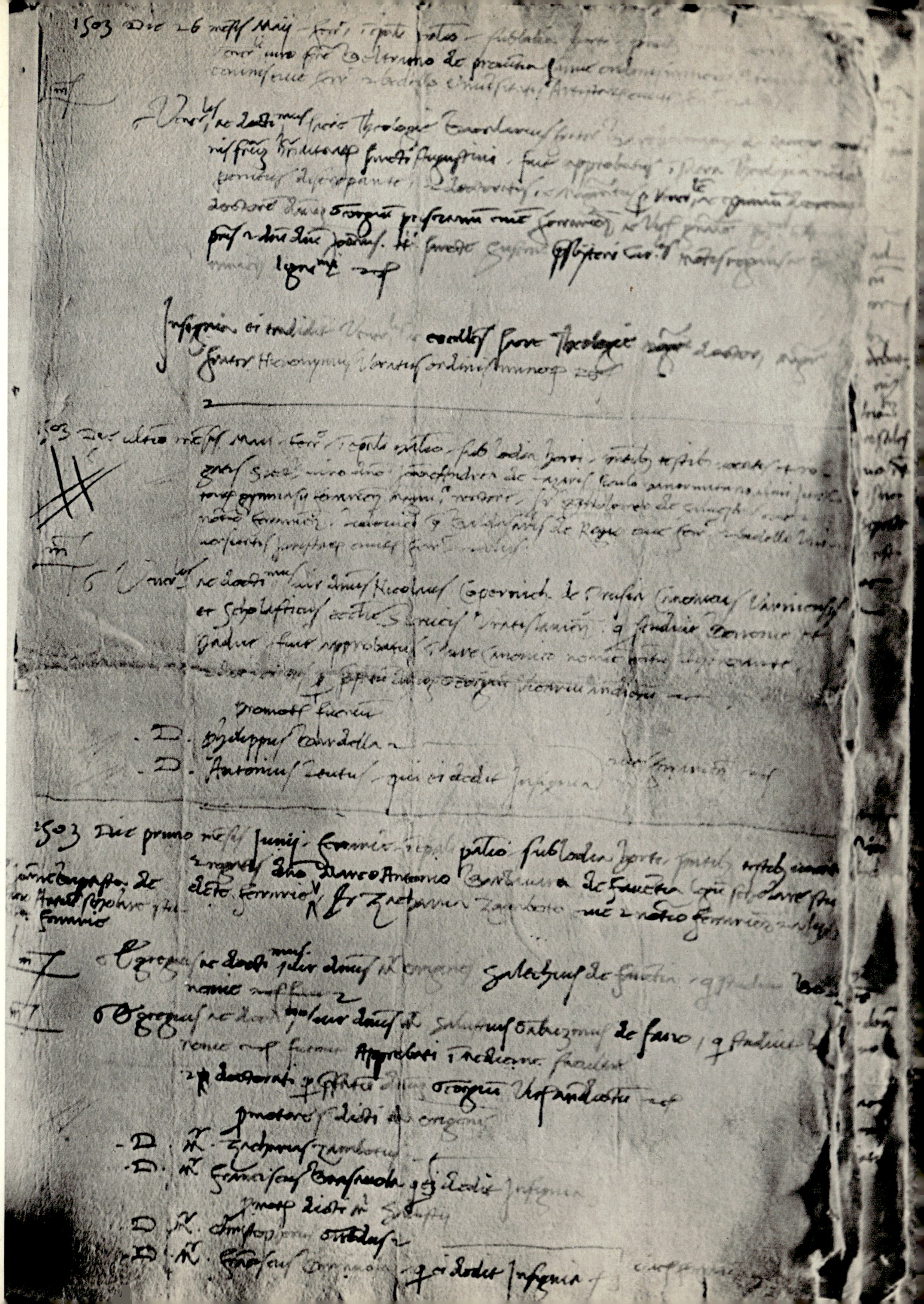

1503 die 26 mensis Maii Ferrariae in episcopali palatio sub lodia horti ... [illegible]

Venerabilis et doctissimus sacrae theologiae baccalaureus ... [illegible] fuit approbatus in sacra theologia ... [illegible]

Insignia ei tradidit ... [illegible] sacrae theologiae doctor ... [illegible]

1503 die ultimo mensis Maii Ferrariae in episcopali palatio sub lodia horti, praesentibus testibus vocatis et rogatis ... [illegible]

Venerabilis et doctissimus vir dominus Nicolaus Copernich de Prusia canonicus Varmiensis et scholasticus ecclesiae Sancte Crucis Vratislaviensis, qui studuit Bononiae et Paduae, fuit approbatus in iure canonico nemine penitus discrepante ... [illegible]

Promotores fuerunt

- D. Philippus Bardella
- D. Antonius Leutus, qui ei dedit insignia

1503 die primo mensis Junii Ferrariae in episcopali palatio sub lodia horti ... [illegible]

... [illegible] Salvatius ... de Fano ... [illegible]

Promotores ... [illegible]

- D. M. Zacharias Zambotus
- D. M. Franciscus ... [illegible] qui ei dedit insignia
- D. M. ... [illegible]
- D. M. ... [illegible] qui ei dedit insignia

◀ Ferrara. Old University Library.

Ferrara. Plaque in the University Library commemorating Nicholas Copernicus.

# WARMIAN LIDZBARK

Lidzbark. View of the city after a copperplate in Jan Krzysztof Hartknoch's "Prussia: Old and New," 1684.

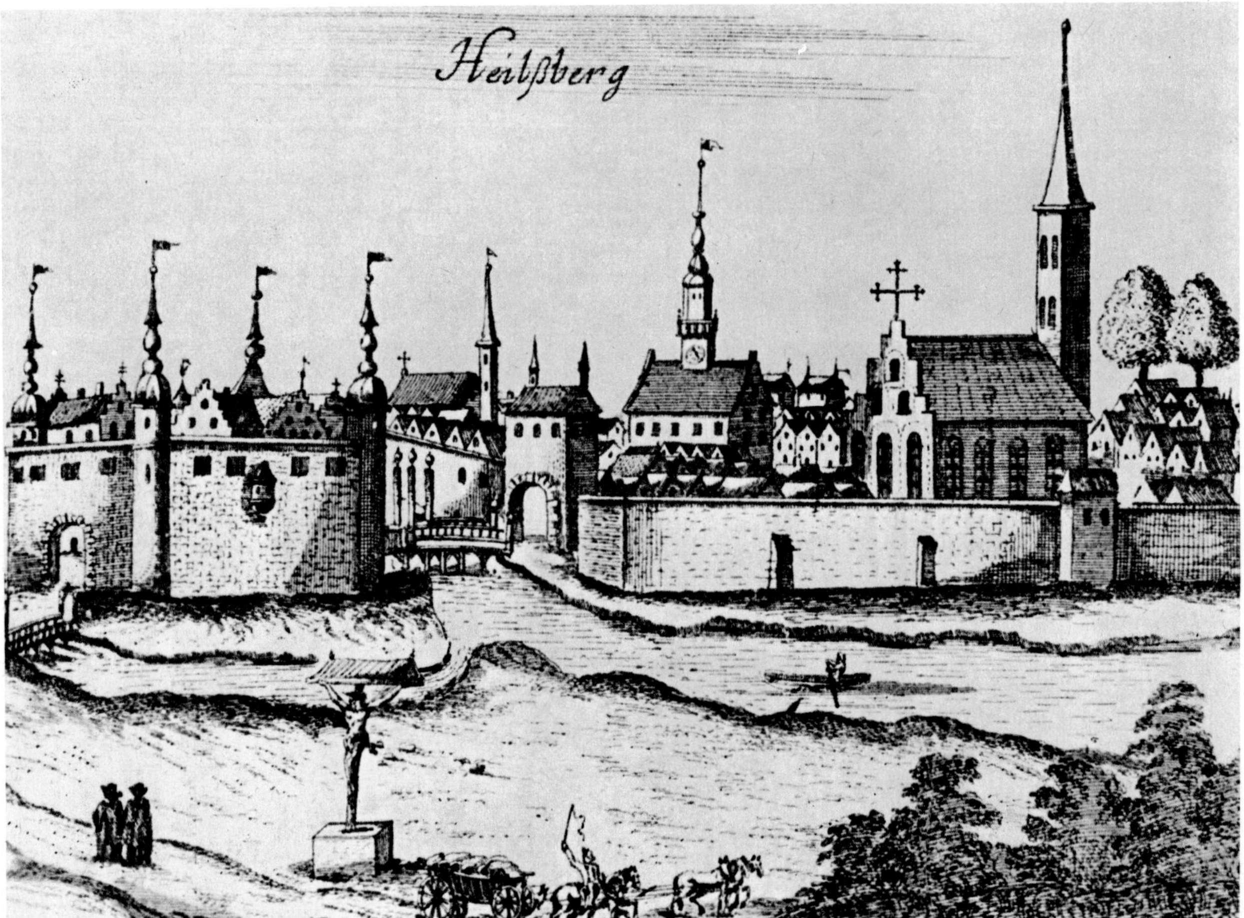

Lidzbark. Castle dating from the second half of the 14th century, the seat of the Warmian Bishops. View from the north-east.

Lidzbark. View of the castle from the south.

Lidzbark. The castle courtyard.

◀ Lidzbark. The arcaded courtyard in the castle dating from the second half of the 14th century.

Lidzbark. Painted frieze showing the coats-of-arms of the Warmian Bishops Łukasz Watzenrode and Fabian Luzjański. ▶

Lidzbark. Part of the exhibition of glass in the ground floor Hall of the castle.

NICOLAUS
LUCAS
FABIANUS
DE LUSIANII
NOBILIS PRUTHENUS
OBIIT
DIE 30. JANUARIS
SEPULTUS
IN CATHEDRALI

Portrait of Tiedemann Bartholomew Giese (1480—1550); Bishop of Warmia from 1549, Nicholas Copernicus' friend.

Portrait of Łukasz Watzenrode (? — 1512), Bishop of Warmia from 1489, uncle and guardian of Nicholas Copernicus.

Title page of poetic letters of Teofilact Symokatta, translated from the Greek into Latin by Nicholas Copernicus and printed by Jan Haller in Cracow in 1509.

Jan Dantyszek (1485—1548), Bishop of Warmia, poet, diplomat, secretary to the King Sigismund the Old.

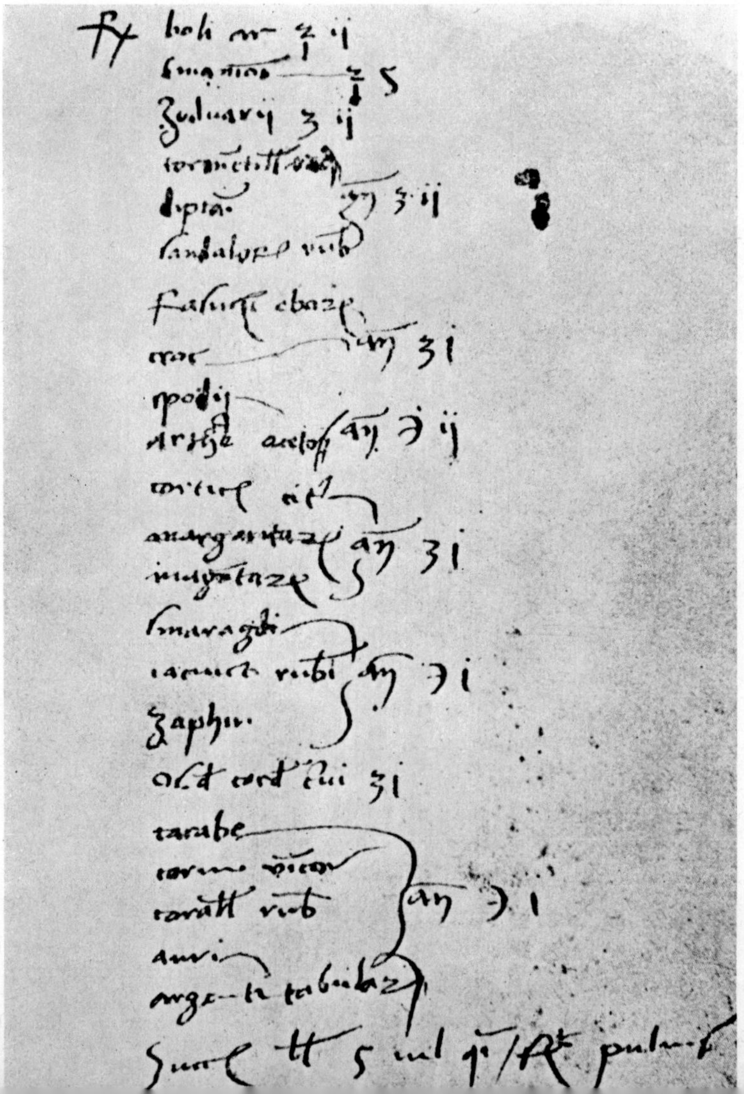

A prescription written by Nicholas Copernicus.

Lidzbark. Part of the castle's historical exhibition.

Lidzbark. The High Gate, dating from the first half of the 14th century.

# ELBLĄG

Elbląg. View of the town from the River Nogat — in the distance the steeple of the church of St. Nicholas dating from the second half of the 13th century, surmounted by a Renaissance dome.

Elbląg. Part of the high altar showing the Adoration of the Three Magi, dating from c. 1520—in the church of St. Nicholas.

Elblag. Dominican Church — dating from the 13th and 14th centuries — east front.

Elbląg. Market Gate, dating from the 14th century, added to in 15th century, and reconstructed after the Second World War.

# MALBORK

Malbork. 14th century Gothic city gate (copestone of a later date).

◀

Malbork. Castle of the Teutonic Knights, built from the 13th to the 15th centuries, renovated in 19th century, rebuilt after a fire in 1958—1964 — south view of the High Castle.

Malbork. View from a window at the courtyard of the High Castle. ▶

Malbork. Refectory in the High Castle. ▶

Malbork. Early castle kitchen.

Malbork. Part of the Middle Castle, dating from the 14th century, seat of the Master of the Order of the Teutonic Knights.

Malbork. View of the courtyard from the gate of the Middle Castle.

Portrait of Prince Albrecht Hohenzollern, the last Grand Master of the Teutonic Knights.

Malbork. Part of the Knights' Hall in the Middle Castle.

# GDAŃSK

Gdańsk. View and map of the town, dating from 1739.

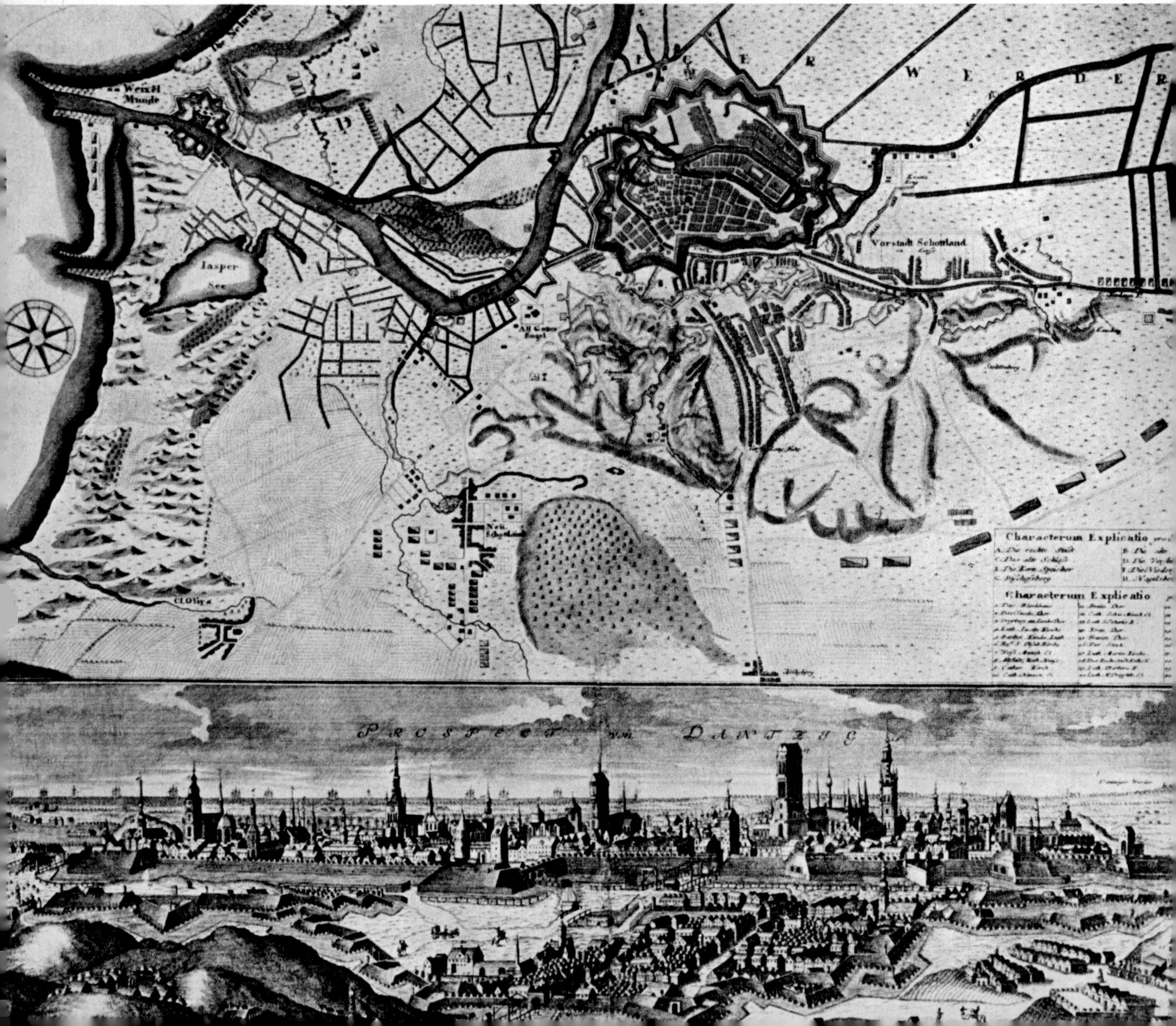

Gdańsk. View of the waterfront with the Mariacka Gate, built in the late 15th century, with an added Renaissance dwelling-house.

Gdańsk. Part of Długa Street with the Złota Gate built in 1612—1614 by the town architect, Abraham van den Blocke, and the 15th century Prison Tower. ▶

Gdańsk. 15th century crane on the Długie Pobrzeże.

Gdańsk. Clock on the corner tower of the Main Town Hall.

Gdańsk. Part of Długa Street showing the Main Town Hall, built 1379—1382 with a late 15th century tower surmounted by a Renaissance dome and a statue of King Sigismund August. ▶

Gdańsk. Façade of the 15th century Artus Court after its reconstruction in 1616—1617 after the design of Abraham van den Blocke (restored after the Second World War).

Gdańsk. 17th century Neptune Fountain in front of the Artus Court.

Gdańsk. Houses lining the Długi Targ after their reconstruction in 1952–1962. ▶

Gdańsk. A late Renaissance portal to a dwelling-house near the Artus Court.

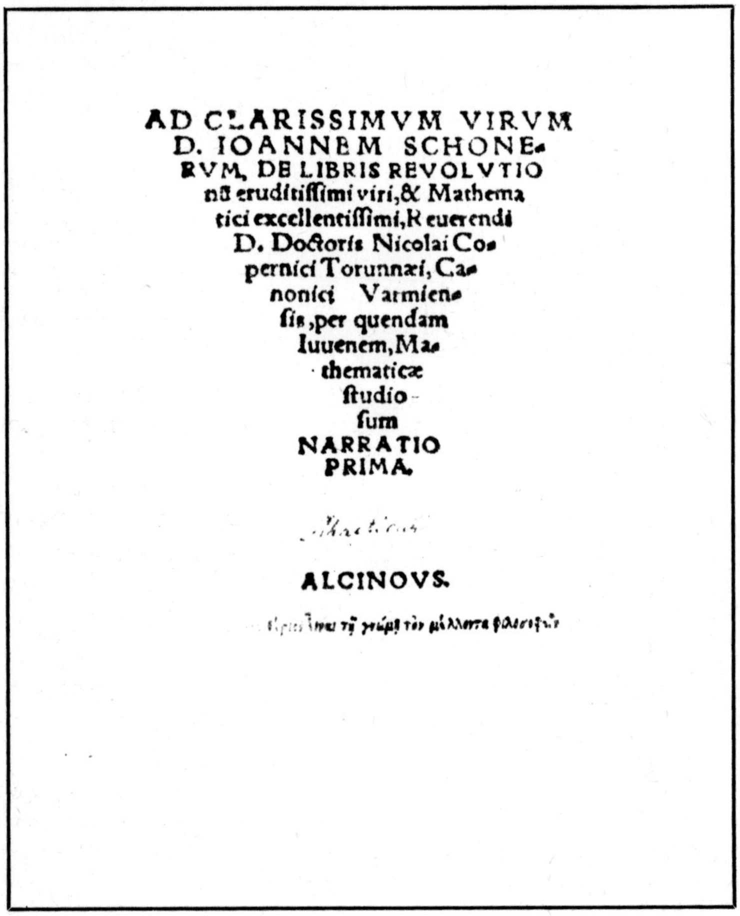

AD CLARISSIMVM VIRVM
D. IOANNEM SCHONE-
RVM, DE LIBRIS REVOLVTIO
nū eruditiſſimi viri, & Mathema
tici excellentiſſimi, Reuerendi
D. Doctoris Nicolai Co-
pernici Torunnæi, Ca-
nonici Varmien-
ſis, per quendam
Iuuenem, Ma-
thematicæ
ſtudio-
ſum
NARRATIO
PRIMA.

ALCINOVS.

[illegible] τῇ γνώμῃ τὸν μέλλοντα φιλοσοφεῖν

Gdańsk. "Narratio Prima" Joachim Rheticus' summary of Nicholas Copernicus' heliocentric theory, printed in Gdańsk in 1540.

CZEKOLADY

Gdańsk. Reconstructed façades of the buildings near Długi Targ, in the distance 16th century Zielona Gate.

Gdańsk. 16th century defensive towers known as the Stągwie Mleczne.

# GRUDZIĄDZ

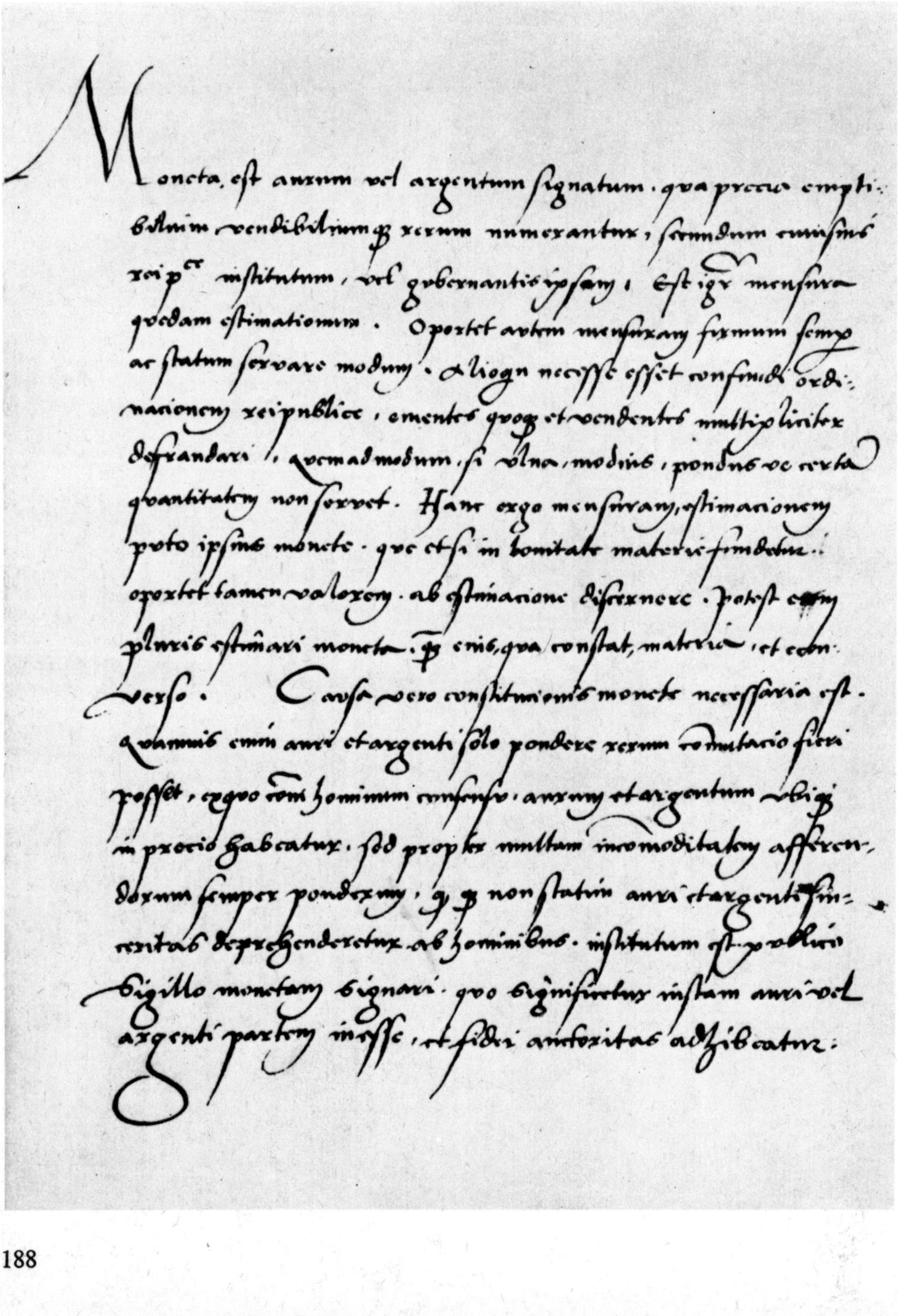
Moneta est aurum vel argentum signatum, qua precio empti-
bilium vendibiliumque rerum numerantur, secundum cuiusque
reipublice institutum, vel gubernantis ipsam. Est igitur mensura
quedam estimationum. Oportet autem mensuram firmum semper
ac statum servare modum. Alioquin necesse esset confundi ordi-
nationem reipublice, ementes quoque et vendentes multipliciter
defraudari, quemadmodum, si ulna, modius, pondus ve certam
quantitatem non servet. Hanc ergo mensuram, estimationem
puto ipsius monete. que etsi in bonitate materie fundetur,
oportet tamen valorem ab estimatione differre. potest enim
pluris estimari moneta quam eius, qua constat, materia, et con-
verso. Causa vero constitutionis monete necessaria est.
Quamvis enim auri et argenti solo pondere rerum commutatio fieri
posset, ergo communi hominum consensu aurum et argentum ubique
in precio habeatur. sed propter multam incommoditatem afferen-
dorum semper ponderum, et quia non statim auri et argenti sin-
ceritas deprehenderetur ab hominibus, institutum est publico
sigillo monetam signari, quo significetur iustam auri vel
argenti partem inesse, et fidei autoritas adhibeatur.

First page of Nicholas Copernicus' treatise: "Tractatus de monetis," copy dating from 1517.

Grudziądz. Plaque on a building near the Market commemorating Nicholas Copernicus' delivery of his treatise on money.

Nicholas Copernicus presents his views on monetary matters in the Town Hall at Grudziądz in 1552—after a drawing by Feliks Sypniewski, c. 1876.

◀ Coins used in Nicholas Copernicus' time.

Grudziądz. Medieval granaries seen from the River Vistula.

Grudziądz. 13th/14th century parish church.

Grudziądz. Medieval buildings with parts of defence walls and the Wodna Gate.

# OLSZTYN

Olsztyn. Castle of the Warmian Chapter built in the late 14th century.

Olsztyn. Glimpse of the castle through the trees on River Łyna.

Olsztyn. The castle courtyard.

Olsztyn. South-east side of the castle and defence walls.

Olsztyn. Reflective gnomonic plaque dating from the first quarter of the 16th century, in the castle gallery. The presumed work of Nicholas Copernicus.

Olsztyn. Cloister in the south-west wing of the castle.

Olsztyn. Keystone with the coats-of-arms of the Warmian Bishops.

Letter of the Warmian Chapter to the Polish King Sigismund I, written by Nicholas Copernicus in 1520, asking for help in defending Olsztyn against the Teutonic Knights.

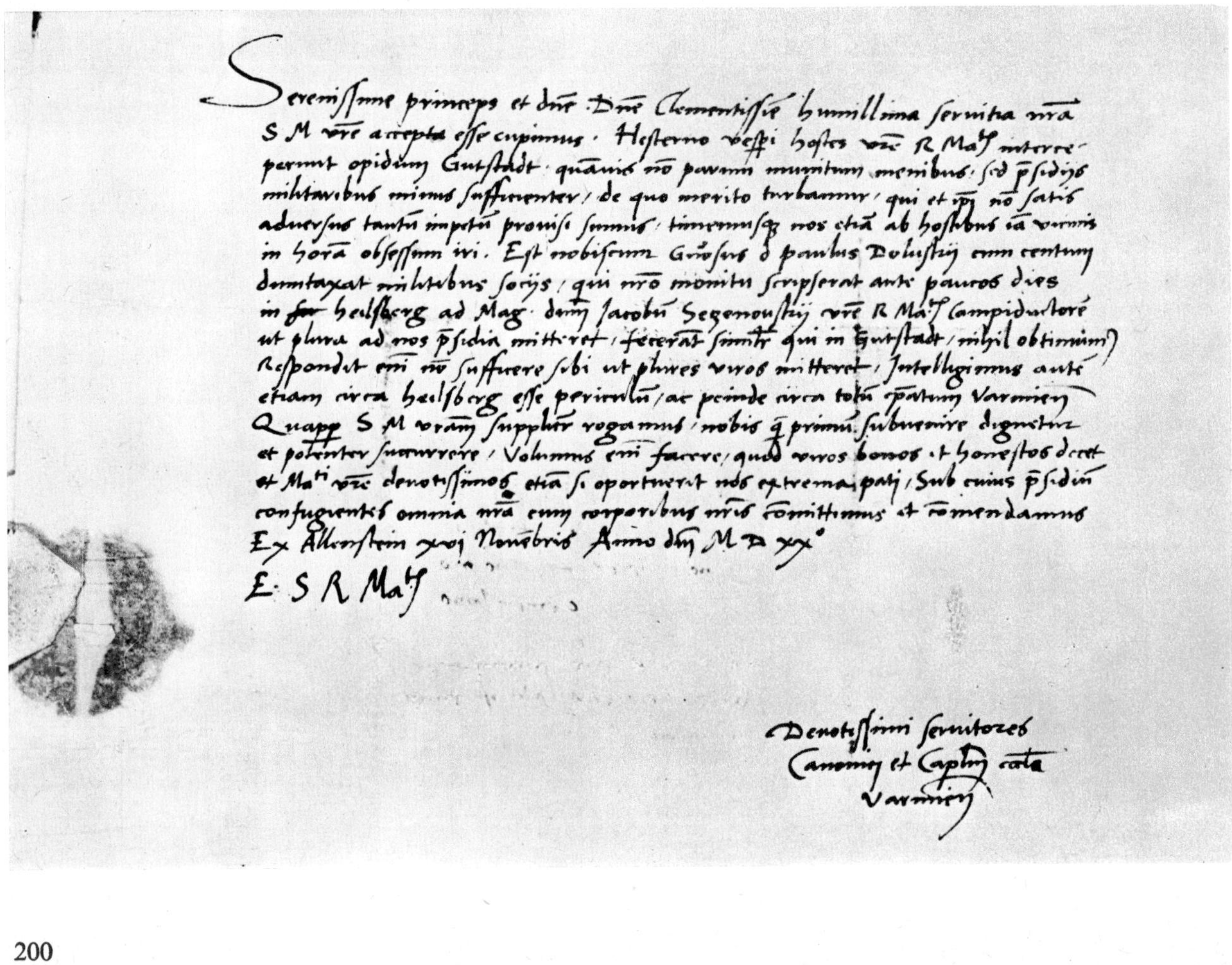

Serenissime princeps et domine Domine Clementissime humillima servitia nostra
S M vestre accepta esse cupimus. Hesterno vespri hostes vestre R Maiestatis intercepe-
runt opidum Gutstadt quamvis non parum munitum menibus, sed presidiis
militaribus minus sufficienter, de quo merito turbamur, qui et ipsi non satis
adversus tantum impetum provisi sumus, timemusque nos etiam ab hostibus in brevis
in hora obsessum iri. Est nobiscum Generosus d Paulus Dolusky cum centum
dumtaxat militibus sociis, qui nostro monitu scripserat ante paucos dies
in Heilsberg ad Mag. dominum Jacobum Segenovsky vestre R Mtis Campiductore
ut plura ad nos presidia mitteret, fecerat similiter qui in Gutstadt, nihil obtinuimus
Respondet enim non sufficere sibi ut plures viros mitteret. Intelligimus autem
etiam circa Heilsberg esse periculum, ac proinde circa totum spatium Varmien.
Quapropter S M vestram suppliciter rogamus, nobis quam primum subvenire dignetur
et potenter succurrere. Volumus enim facere, quod viros bonos et honestos decet
et Mti vestre devotissimos etiam si oportuerit nos extrema pati, Sub cuius presidium
confugientes omnia nostra cum corporibus nostris committimus et commendamus
Ex Allenstein xvi Novembris Anno dni M D xx°
E. S R Mtis

Devotissimi servitores
Canonici et Capitulum ecclesie
Varmien.

The Polish King Sigismund I, the Old.

Document concerning untilled fiefs, compiled by Nicholas Copernicus in his capacity as administrator of the Chapter treasury in the Pieniężno districts (1517—1519) and Olsztyn (1516—1521).

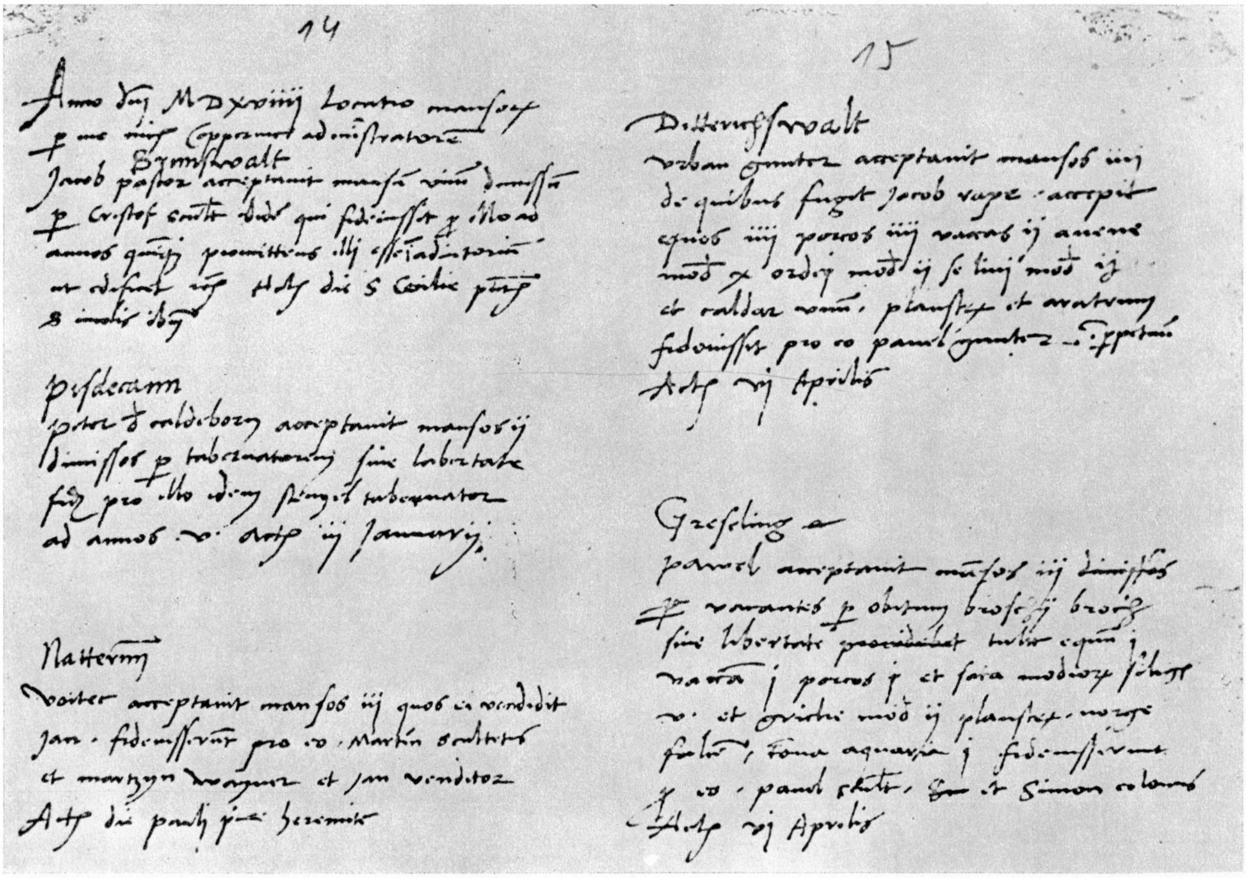

Olsztyn. Part of the Knights' Hall on the first floor of the castle.

Olsztyn. Plaque on the plinth of the statue.

Olsztyn. Statue of Nicholas Copernicus in the square before the castle — view from the west.

Olsztyn. Town Hall in the Old Town, rebuilt after the Second World War.

Olsztyn. Colonnades of historic buildings in the Market Place.

Nicholas Copernicus — 16th century portrait called "Gołuchowski," after the place it was kept for long time, lost during Second World War hostilities.

A typical Warmian landscape. ▶

Olsztyn. Wysoka Gate, dating from the 14th century — the last remnants of the town's defence walls.

# FROMBORK

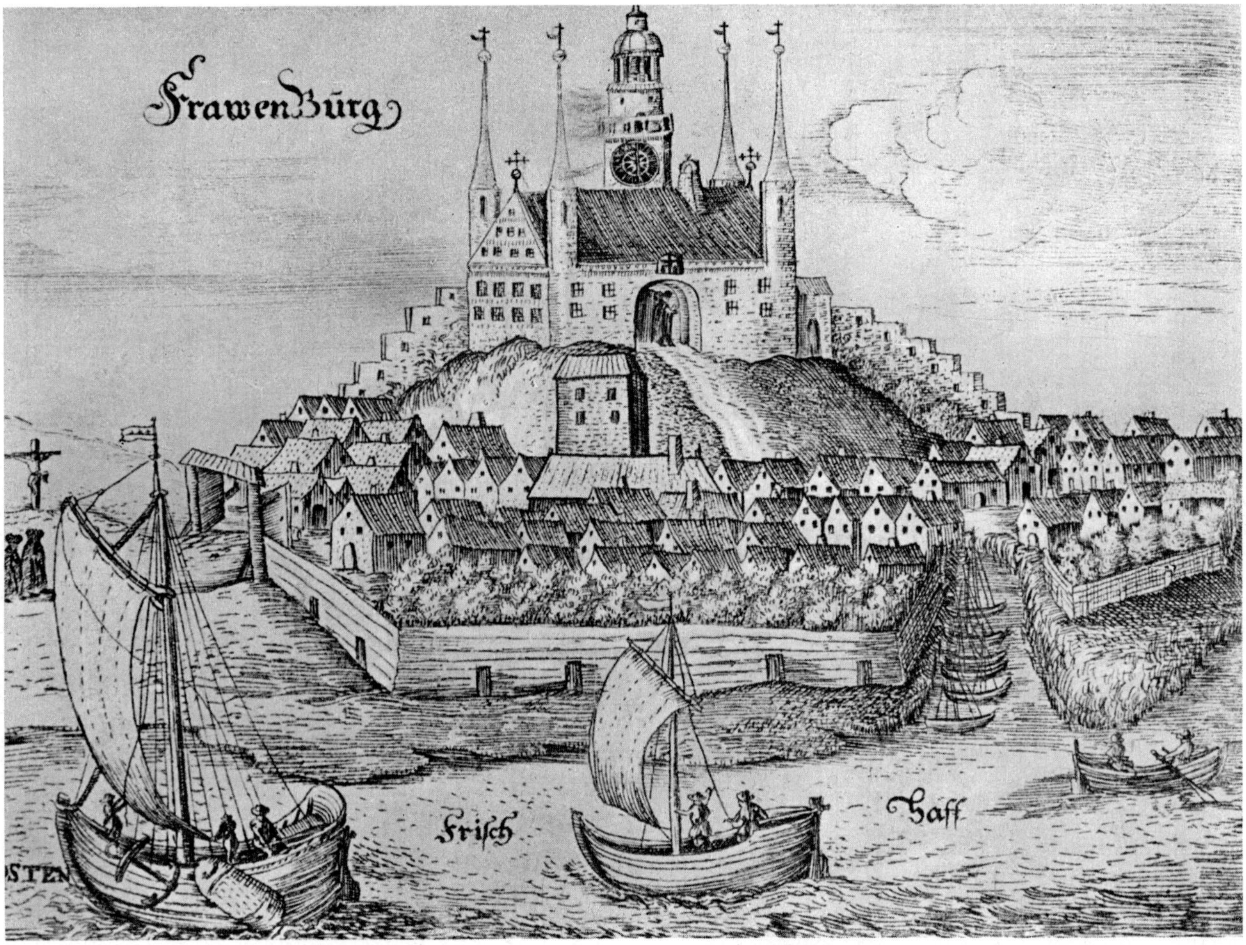


Frombork. The town seen from the Vistula Bay — after a copperplate in Jan K. Hartknoch's "Prussia: Old and New," 1684.

Frombork. Part of the 14th century ring of defence walls, showing the tower gate and so called Copernicus Tower on Cathedral Hill.

Frombork. Part of the 14th century defence walls between the Cathedral and the North Tower.

Frombork. The tower at the Main Gate, seen from the South Tower.

Frombork. Part of the defence walls, seen from the South Bastion. In the distance — the Cathedral spire.

Frombork. Cathedral — built in 1329–1388, seen from the north-west.

Frombork. Epitaph to Nicholas Copernicus, donated by the Warmian Chapter c. 1735 and placed in the Nave of the Cathedral.

Frombork. Cathedral — view of the Nave with the 14th century starshaped dome and the Baroque interior.

◀ Frombork. Late-Gothic polyptych, dating from 1504 (at one time the high altar) in the north nave of the Cathedral.

Frombork. The Baroque altar of St. Bartholomew, known as the Copernicus' Altar.

Frombork. Building of the Canonical Curia — the original location of the Copernicus' Museum from 1948 to 1967.

Frombork. The Gothic-Baroque Bishop's Palace, known as the Ferber's Palace—built in the 16th century, reconstructed in 1727—houses now the Copernicus' Museum.

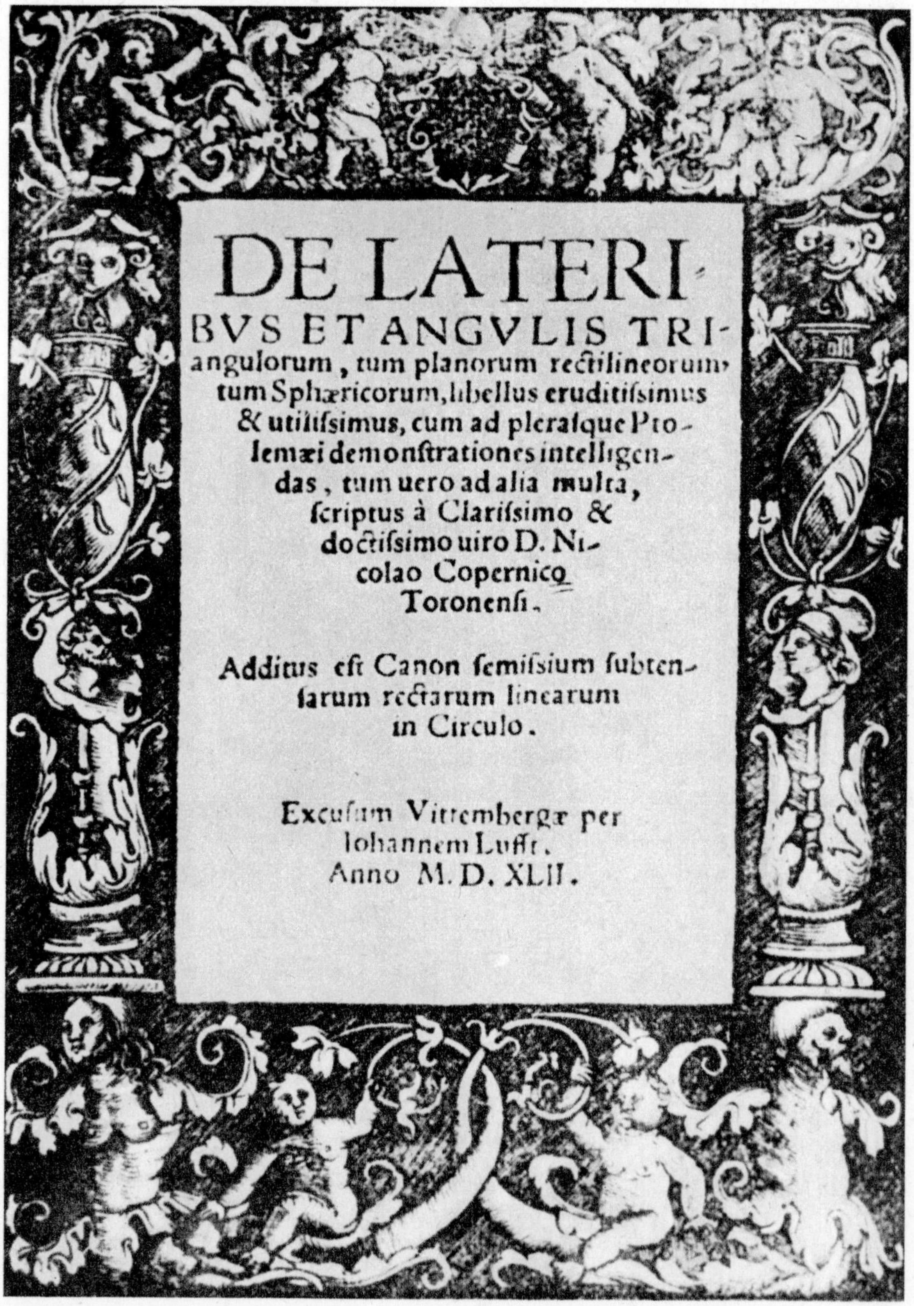
DE LATERI-
BVS ET ANGVLIS TRI-
angulorum, tum planorum rectilineorum,
tum Sphæricorum, libellus eruditissimus
& utilissimus, cum ad pleraſque Pto-
lemæi demonſtrationes intelligen-
das, tum uero ad alia multa,
ſcriptus à Clariſsimo &
doctiſsimo uiro D. Ni-
colao Copernico
Toronenſi.

Additus eſt Canon ſemiſsium ſubten-
ſarum rectarum linearum
in Circulo.

Excuſum Vittembergæ per
Iohannem Lufft.
Anno M. D. XLII.

Title page of Nicholas Copernicus' "Treatise on Trigonometry," Wittenberg in 1542.

◀ Nicholas Copernicus library, now housed in Uppsala, Sweden.

Pope Paul III (1468—1549), to whom Copernicus dedicated his "De Revolutionibus" — after a 16th century copperplate.

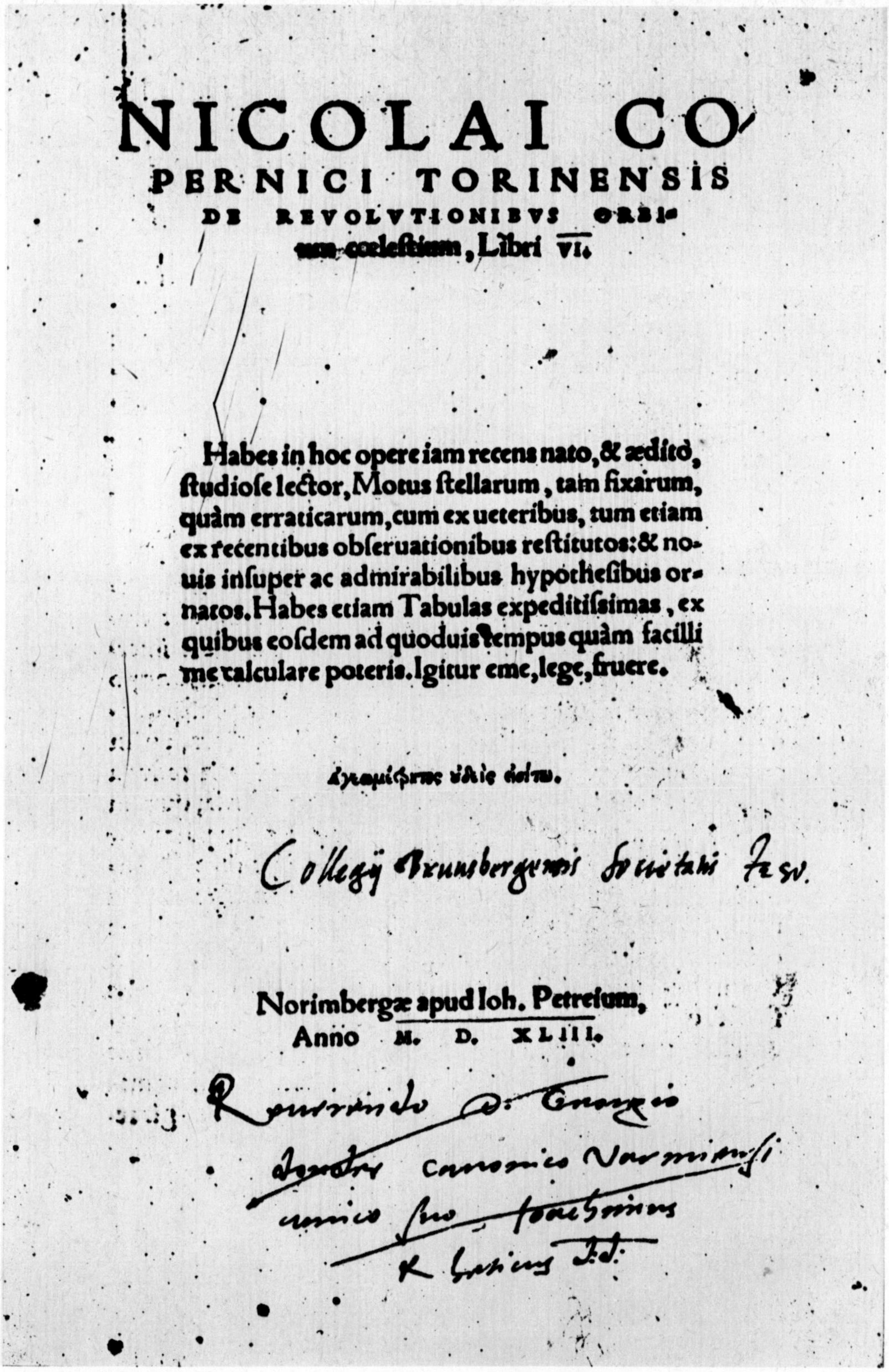

NICOLAI CO
PERNICI TORINENSIS
DE REVOLVTIONIBVS ORBI-
um cœlestium, Libri VI.

Habes in hoc opere iam recens nato, & ædito, studiose lector, Motus stellarum, tam fixarum, quàm erraticarum, cum ex ueteribus, tum etiam ex recentibus obseruationibus restitutos: & nouis insuper ac admirabilibus hypothesibus ornatos. Habes etiam Tabulas expeditissimas, ex quibus eosdem ad quoduis tempus quàm facillime calculare poteris. Igitur eme, lege, fruere.

ἀγεωμέτρητος οὐδεὶς εἰσίτω.

Collegij Brunsbergensis Societatis Jesu.

Norimbergæ apud Ioh. Petreium,
Anno M. D. XLIII.

Reverendo D. Georgio
canonico Varmiensi
amico suo

Title page of the first edition of Copernicus' work "De Revolutionibus," printed in Nuremberg in 1543.

Frombork. Copernicus' Museum — Astrolabe against the background of a copy of Jan Matejko's painting.

Frombork. Copernicus' Museum — part of the exhibition.

Frombork. Statue commemorating Nicholas Copernicus, donated by Scouts in 1971.

Frombork. Copernicus' Museum — illustration of "Polish Sky' after a 17th century Atlas.

Corn fields by the Warmian lakes. ▶

Sunset in Warmia.

# COPERNICUS' DISCIPLES

Title page of Galileo's work "On two structures of the World" (1641), on the left Galileo Galilei (1564—1642) after a 15th century copperplate.

The German astronomer and mathematician Johannes Kepler (1571—1630), who developed Copernicus' conception — after a 17th century copperplate.

Part of the Astronomical Observatory in Nuremberg, after a drawing of 1648.

Title page of the "Atlas of the Sky" published in Nuremberg in 1741 with portraits of the distinguished astronomers: Claudius Ptolemy (second century AD), Nicholas Copernicus (1473—1543), Johannes Kepler (1571—1630), Tycho de Brahe (1546—1601).

Stylized drawing (1741) illustrating the planetary system.

Juri Gagarin — first man in space.

The first man sets foot on the moon — astronaut Armstrong on the Silver Globe.

Bertel Thorvaldsen's of Nicholas Copernicus in front of the Staszic Palace, enveiled in 1830.

# REFERENCES

In writing this book I used numerous references, however allow me to list those from which I extracted many quotations:


Nicolaus Copernicus: *De revolutionibus orbium coelestium. Liber Primus,* Polish translation by Mieczysław Brożek. PWN, Warsaw 1953.

Ludwik Antoni Birkenmajer: *Mikołaj Kopernik* (Nicholas Copernicus), Cracow 1900.

Karol Górski: *Dom i środowisko rodzinne Mikołaja Kopernika* (Home and family background of Nicholas Copernicus), PWN, Toruń 1968.

Hermann Kesten: *Copernicus und seine Welt,* Polish translation by Krzysztof Radziwiłł and Janina Zeltzer, PIW, Warsaw 1961.

*Komentarze fromborskie* (Frombork Commentaries), Periodicals published by "Pojezierze," Olsztyn 1965, 1968.

*Mikołaj Kopernik. Szkice monograficzne* (Nicholas Copernicus. Monographic Studies), PWN, Warsaw 1965.

Thomas S. Kuhn: *The Copernican Revolution,* Harvard University Press, Cambridge 1957, Polish translation by Stefan Amsterdamski, PWN, Warsaw 1966.

Riccardo Lewański: *Storia delle relazioni fra la Polonia e Bologna,* Bologna 1951.

Stanisław Łempicki: *Renesans i humanizm w Polsce* (Renaissance and Humanism in Poland), Czytelnik, Warsaw 1952.

Jerzy Sikorski: *Mikołaj Kopernik na Warmii* (Nicholas Copernicus in Warmia), Research Station of the Polish Historical Society, Olsztyn 1968.

Jeremi Wasiutyński: *Kopernik. Twórca nowego nieba* (Copernicus, Founder of a New Sky), Warsaw 1938.

Vitruvius: *O architekturze ksiąg dziesięć* (Ten Books of Architecture), Polish translation by Kazimierz Kumaniecki, PWN, Warsaw 1956.

Henryk Zins: *W kręgu Mikołaja Kopernika* (In the Sphere of Nicholas Copernicus), Wydawnictwo Lubelskie, Lublin 1966.

ZAKŁADY GRAFICZNE
DOM SŁOWA POLSKIEGO — WARSZAWA
ZAM. 82/a/73